GAME CHANGER

GAME CHANGER

The Technoscientific
Revolution in Sports

RAYVON FOUCHÉ

JOHNS HOPKINS UNIVERSITY PRESS   BALTIMORE

Johns Hopkins University Press
2715 North Charles Street
Baltimore, Maryland 21218-4363
www.press.jhu.edu

Library of Congress Cataloging-in-Publication Data

Names: Fouché, Rayvon, 1969– author.
Title: Game changer : the technoscientific revolution in sports / Rayvon Fouché.
Description: Baltimore : Johns Hopkins University Press, [2017] | Includes
    bibliographical references and index.
Identifiers: LCCN 2016028462 | ISBN 9781421421797 (hardcover : alk. paper) |
    ISBN 9781421421810 (electronic) | ISBN 1421421798 (hardcover : alk. paper) |
    ISBN 142142181X (electronic)
Subjects: LCSH: Sports—Technological innovations. | Sports—Physiological
    aspects. | Performance technology. | Doping in sports. | Sports sciences.
Classification: LCC GV745 .F68 2017 | DDC 688.7/6—dc23
LC record available at https://lccn.loc.gov/2016028462

A catalog record for this book is available from the British Library.

*Special discounts are available for bulk purchases of this book. For more information,
please contact Special Sales at 410-516-6936 or specialsales@press.jhu.edu.*

Johns Hopkins University Press uses environmentally friendly book materials,
including recycled text paper that is composed of at least 30 percent post-consumer
waste, whenever possible.

Contents

Acknowledgments   *vii*

INTRODUCTION. Sports, Bodies, and Technoscience   1

PART ONE: JUDGING ARTIFACTS

1   Black Is the New Fast: Swimsuit Technoscience and the
    Recalibration of Elite Swimming   31
2   Gearing Up for the Game: Equipment as a Shaper of Sport   67
3   Disabled, Superabled, or Normal? Oscar Pistorius and
    Physical Augmentation   100

PART TWO: EVALUATING BODIES

4   "I Know One When I See One": Sport and Sex Identification in an
    Age of Gender Mutability   131
5   The Parable of a Cancer Jesus: Lance Armstrong and the Failure of
    Direct Drug Testing   154
6   "May I See Your Passport?" The Athlete Biological Passport as a
    Technology of Control   178

CONCLUSION. Body, Motor, Machine: The Future of Technology
    and Sport   205

Notes   *223*
Index   *255*

Acknowledgments

Books often have convoluted origin stories. This book began some time back in the early 1990s on rural roads in central Illinois guarded by tall corn and much shorter soybeans. Thousands of miles of suffering in silence, through wind, rain, heat, and snow with the same small group of people, gave me plenty of time to wonder about the world in unusual ways. Though it took many years to return to technology and sport, I have always been interested in the pair.

At the conclusion of this writing, I have more intellectual debts than I can ever attempt to repay. Much of the book and most of my scholarly work seeks to think through Langdon Winner's timeless question, do artifacts have politics? For me, his question has morphed into: (1) How do objects create social and cultural meaning for individuals and communities? and (2) How do people endow and embed social and cultural meaning within objects? His question is still fresh and exciting. Thanks, Langdon.

I presented portions of this project at a host of venues. A heartfelt thanks goes to the audiences at Boston University, Carnegie Mellon University, the Chicago Humanities Festival, Cornell University, the IDSA Central District Design Conference, Kendall College of Art and Design, Morrisville State College, National University of Defense Technology (Changsha, China), Osaka City University (Osaka, Japan), Purdue University, Tsinghua University (Beijing, China), University of California–Berkeley, University of Illinois–Urbana-Champaign, and University of Michigan. Their engaged comments profoundly shaped this book, and I will be forever grateful.

As I drafted this book, I have had the pleasure to research and teach at the Rensselaer Polytechnic Institute (science and technology studies), the University of Illinois (Center for East Asian and Pacific Studies, History, and Information Trust Institute), and most recently Purdue University as director of the American Studies program in the School of Interdisciplinary Studies. In each place, I found kind, generous, and thoughtful students, faculty, and staff who made this entire endeavor possible. I extend my deepest gratitude to you all.

I have been fortunate to have supportive friends, colleagues, and family members spanning all parts of the globe. They have always embraced the perplexing questions I ask of the world. I am indebted to you all.

Finally, I thank my wife, Sharra Vostral, and my son, Eads. They have both lived every word on each page. Their presence continually makes life extraordinarily beautiful. Thank you for your time and patience. Much love to you both.

GAME CHANGER

Sports, Bodies, and Technoscience

"It's the motor, not the machine." As a competitive cyclist in the late 1980s through the mid-1990s, I came to speak this phrase regularly. It became a familiar mantra that my compatriots and I would systematically repeat. Put simply, this saying means that one's body—the motor—is vastly more important than any technoscientific device, pharmaceutical treatment, or psychological conditioning in the final outcome of a cycling race. In hindsight, I repeated this phrase to others and myself out of my own hubris, ignorance, and, most importantly, denial. This overly self-confident utterance by an aspiring twentysomething athlete should not come as a shock because a large part of elite competitive sport is about sustaining an unwavering, and often illogical, belief in oneself. It is a belief in the superiority and infallibility of one's body that undergirds this way of thinking. The body is potentially the only aspect of an athletic competition that an athlete can control completely. When the "game" gets tough, the body and the mental and physical training absorbed by it will, one hopes, carry one through to triumph. This ideal centers on the belief that when all else fails, the "motor" will transcend all and deliver an athlete to victory. Thus, it should not come as a surprise that the phrase "it's the motor, not the machine" mollified insecurities, inspired confidence, and supplied motivation to carry on, even when chances for success grew ever slimmer. The growing popularity of this phrase among cyclists, and others similar to it in other sports, stems from anxieties about uncontrollable unknowns.[1]

One of the major aims of this book is to investigate how sporting communities respond to the uncontrollable unknowns that emerge from the inseparable interweaving of science and technology, or, more accurately, technoscience. In my case, as much as I attempted to push the technoscientific out of my consciousness, I was fully aware that technoscientific objects, practices, and procedures played a role in the final outcomes of races. By the 1990s, cycling, like most other sports, experienced a technoscientific explosion ranging from lightweight composite materials like carbon fiber to stretchy wind-cheating fabrics like elastane. Also during this period, sport began to embrace the scientific rigor that physicians, psychologists, and scientists brought to training, nutrition, and competitive preparation. These technoscientific pathways—just like in the broader society—progressed at a quickening and uncomfortable pace. Athletes like myself began to wonder whether these technoscientific evolutions would become revolutions, with the technoscientific overshadowing the corporeal to a point where athletic competitions would no longer be between athletes on the field, but between scientists and technologists in the lab.

As rhetoric around the "lab" in sport developed over the past few decades, the "lab" that dominated public and media dialogues was not the engineering lab where mechanical engineers and material scientists developed new sporting equipment, but the lab where biochemists and pharmaceutical engineers manipulated human cells to extract maximum performance. It is in this regard that the use, potency, and legality of performance-enhancing drugs have come to dominate discussions and debates about technoscience and sport. Unfortunately, for many, technoscience in sport has become synonymous with performance-enhancing drugs.

Until the last few decades of the twentieth century, the idea that athletes would use *any* object or substance—regardless of its legality or its impact on future health—to improve their chances of winning was a foreign concept for a large portion of sport's viewing publics. These publics believed, at least in an American context, that elite athletes who willingly chose to use a questionable enhancing device or substance were rare. Only unethical or rogue athletes would stoop so low as to demean themselves and their sports, or irreparably damage their bodies, by "cheating." But as recent admissions and exposés in every major and minor market sport around the world have shown, the use of substances that athletes, trainers, or random acquaintances believed would improve performances and the ability to win are as old as sport itself. Even casual viewers of sport now understand that the late twentieth- and early twenty-first-century eras have become defined by lean mass–building substances such as human growth hormone (HGH) and blood-boosting drugs such as erythropoietin (EPO). The ubiquity of performance-enhancing substances has led to the dominant perception that the most prominent technoscientific changes within sport have been pharmaceutical.

Though the medical revolutions shaping the broader world unquestionably have reconfigured sport, the past fifty years of sporting history have seen many other less publicly volatile technoscientific developments profoundly influence the games people play. We, as a society, have given great attention to the drugs that influence sport, but the material artifacts that reconfigure the games we love to watch and play generally have not been a cause for alarm or investigation. This prominent focus on performance-enhancing substances obfuscates other technoscientific forms that impact sporting competitions. In order to gain a broader understanding of the ways in which science and technology—for better or worse—transform sport, this book aims to invert the weight of the relationship between performance-enhancing drugs and all other technoscience. By broadening the focus of science, technology, and sport beyond drugs and conceptually recalibrating this relationship, other increasingly important technoscientific artifactual interactions will come into higher relief. Extending discussions of sport beyond the well-trodden ground of performance-enhancing drugs will allow publics, athletes, governing institutions, engineers, scientists, and designers to gather a more contextualized understanding of the multiple ways in which technoscientific products have shaped and will shape the landscape of sport.

The current relationship between athletes and equipment, bodies and technoscience, or the motor and the machine is messy at best. Every sport has its requisite gear, but historically sports sublimate this gear to individualized performances and the cultural assertion that the athlete, the body, or the motor has been and will always be vastly more important and valuable than the equipment, the technoscience, or the machine. But what happens—and what does it mean for the social and cultural infrastructure of sport—when technoscience reveals itself to be more than the instruments of a game? These disjunctures expose a set of assumptions that privilege the human body to the point that sport regularly disavows the substantive impacts that technoscience has on the outcomes of games people play.

In the past few decades, it has become harder and harder to deny or even ignore the impact of technoscientific equipment on sport. Athletic events from bass fishing to the America's Cup have shattered the illusion that technoscience is just merely equipment and highlights just how dependent sport is on myriad technoscientific artifacts and practices. Yet, for the most part, contemporary society still views sport as a decidedly human physical endeavor. Humans are social creatures, and the historical emphasis on the human motor over technoscientific machinery not only motivates athletes when the scary efficacy of the next technoscientific implement can be seen on multiple horizons but also supports the collective rejection of the present real power of technoscience in sport. Dominant narratives of athletic and sporting competitions are about human physical and intellectual struggle as well as a host of cultural assumptions, beliefs, and practices that work to privilege the human body over the technoscientific.

To sustain the elevated status of the athlete and the human body in these narratives, technoscience must appear as mere instruments of a game to obscure the intricate interplay among people, institutions, rules, regulations, and technoscience. Understanding the place of technoscience within sport goes beyond the tensions between body and machine. Sport is bound together by and through an evolving network of athletes, engineers, designers, publics, and technoscientific artifacts. This interconnected web of human and nonhuman elements comprises an overlapping set of sporting cultures that form around a specific sport or sporting activity. Within these sporting cultures, technoscience exists as an uncomfortable problem that will continue to produce palpable disruptions within these cultural communities if we cannot build a coherent way to evaluate, assess, and understand the roles of technoscience in multiple sporting domains.

Phrases like "it's the motor, not the machine" exist because of the historical tensions between the human body and equipment within sporting competition. Yet this phrase performs different work depending on one's position. For athletes, phrases of this type represent a set of psychological and rhetorical motivational structures designed to inspire athletes that, if they continue to flog their bodies through gut-wrenching training sessions, they will someday become "champions." This blind reliance on and belief in the body is part of what makes sport so seductive, because in multiple instances athletic bodies proved themselves to be otherworldly through miraculous and transcendent performances. Roger Bannister's 1954 sub-four-minute mile at the Iffley Road track at Oxford University is one such moment.[2] This event—and many others, such as Bob Beamon's 8.90 m long jump at the 1968 Mexico City Olympic Games—represents the triumph of the mind and body over the natural world. Bannister's and Beamon's bodies defied wind, weather, gravity, and nature itself to do the seemingly impossible. For athletic competitions such as track and field, there have been historical moments when the human body invariably has been significantly more important than any manmade technoscientific contrivance. It is these sporting occurrences that sustain the power of the body narrative within sport.

Over the past fifty years, the body narrative has begun to show signs of stress fractures. More recently, the increasing power of technoscience has exposed, and in some cases ruptured, the coherence of the body narrative. Instead of understanding what these technoscientific changes mean, sport often closes ranks and attempts to quickly, and often haphazardly, patch these expanding crevasses. From elite-level soccer blocking any form of goal-line technology until the 2014 World Cup as a belated response to the barrage of YouTube-able instant replay videos exposing the limitations of the human eye in determining close goals, to automobile racing's endless tweaking of regulatory formulas to blur engineering potency as a means to reclaim driver skill and ability, sports fans, competitors, governing bodies, and equipment manufacturers are now living in an era where tough decisions will have to be made in order to determine if sport, in the future, will be more about the motor, the machine, or some transient equilibrium between these two poles.

Thus, the central question this series of studies will address is: In a world defined by its technoscientific output, how does technoscience influence, shape, and challenge the ways societies play, experience, and consume the multiple manifestations of sporting competitions? To reveal these tensions and explore what is at stake for the future evolution of sport, the following case studies examine the complex and convoluted ways sporting cultures maintain the primacy of the body by purposely and inadvertently downplaying or derailing the power of technoscientific artifacts. By highlighting sport technoscience and its impact on athletic competition, this book inherently creates a dialogue that destabilizes the primacy of the body within sporting narratives as well as reveals the importance of technoscience for the stability, sustenance, and maintenance of sporting cultures.

Technoscience and Sport Meritocracy

Over the past fifty years, sport has undergone such massive technoscientific transformation that fans, athletes, governing bodies, and equipment manufacturers struggle to manage the technoscientific sporting worlds they have created. In a compartmentalized and simplified world, equipment manufacturers create and athletes use new technologies that test the limits and boundaries delineated by sport governing bodies while simultaneously upholding the public's expectation of fair and equal competition. Yet we inhabit a world that is far from ideal, and technoscience can breach cherished narratives that validate the primacy of the human body within sporting competitions. Some may believe that sport is about character, masculinity, or a test of physical and mental resolve, but at its core, sport is a social practice and cultural activity. On the surface, it seems as if these interpretations have very little, if anything, to do with technoscience. But over the past century, a forceful wave of sporting technoscience has transformed the social and cultural phenomenon of sport from an athletic endeavor to a technoscientific proving ground. The silent and powerful infiltration of technoscience into sports training, competition, and ideology raises a host of important and increasingly pressing questions for the future of athletic competition. Is technoscience a medium through which athletes can express their physical ability, or is it a network of tools used to undermine the tradition of sport? Can technoscience, in the name of character, ethics, and tradition, effectively determine who should and should not fairly compete? Does technoscience create a false sense of security that rule breakers can be monitored and caught?

So how did we get *here*? And where is that *here*? That *here* is a location where a heightened sense of morality and authenticity drive public discourse about sport. That *here* is a site where corporate institutions invest significant sums of money to produce the next game-winning device. That *here* is also a place where certain technoscience is seen as undermining the purity of the game. Leo Marx, in the *Machine in the Garden*, highlights the social, cultural, and psychological shock people experience when modern industrialization intrudes on their pleasant, peaceful, pastoral landscapes.[3] Technology forced society to uncomfortably rethink its relationship with natural-world beauty in an age of mechanized production. In similar ways, technoscience is forcing sport to reassess the place of man-made material artifacts on the competitive field. The sports field was once a pastoral place of athletic leisure but has potentially become a site for expressions of twentieth-century technoscientifically inflected mass individualism.

As Marx implored us to delve into what it means to have the rural countryside altered by technology, I am asking us to consider what is socially and culturally at stake when sport is reshaped by our technoscientific creativity. This tension does seem somewhat ironic in a world powerfully defined by its scientific and technological invention and innovation. In most cultural spaces, technoscience is seen as opening windows to an improved future. Though scholarly and nonacademic critics, from Lewis Mumford to Michael Pollan, have attempted to destabilize these progressive narratives, technoscience is still seen predominantly as a gateway to a "better" future.[4] Over the past twenty years, though, the unbridled enthusiasm for technoscience in sport has started to wane. If one surveys major sports journalism outlets such as *Sports Illustrated*, *Sporting News*, or *ESPN The Magazine* over that period, it is hard to miss the extensiveness of technoscientific treatments, materials, and artifacts that enable athletes to perform in ways that transgress the agreed upon parameters of fair and equal competition. In the same time period, nearly every major sport has participated in rewriting their rulebooks to contain, delimit, or repel some form of technoscientific innovation.[5] In this context, the looming question is, has technoscience gone too far?

What is "too far" depends on the history and tradition of a sport. From where I sit, as a scholar who researches and teaches about society's interactions with technoscience over the past two centuries, the "too far" question has been asked since humans began using technology and science to alter ways of living.[6] It is a fundamental question that lurks behind the assessment of every technoscientific evolution.[7] In the United States, some may see the resuscitation of craft industries such as leatherworking, artisanal foods, and digital marketplaces (e.g., Etsy) as a reaction to the loss felt by the mass production of everything. Those in the handcraft movements would contend that the resurgence of craftsmanship has been conjured up by a desire to return to a bygone era in which society and the material artifacts of that society were not only better and longer lasting but also more pure and authentic.[8] Late twentieth- and early twenty-first-century cultural reevaluations of the mechanization, digitalization, and the eventual Googlization of everything are mirrored in the world of sport.[9] Though critical resistance to technoscience always has been present, many contemporary critics direct their concerns at the perceived loss of sport's history, tradition, culture, and identity.[10] The questioning of technoscience has become even louder as competitions migrate from demonstrations of athletic ability to multilevel-marketed entertainment extravaganzas where every performance needs to be quantified in an effort to understand how that performance adds or subtracts value from a specific athletic enterprise.

Though not writing about technoscience and sport, journalist Howard Bryant captures one aspect of what fans love about sport when he writes that "meritocracy . . . is the basic draw of sports: your best against mine, the scoreboard oblivious to pedigree, race, class or gender. The promise of pure competition is perhaps the biggest reason we watch." Bryant also indicates that this idea is a fantasy. He concludes that sport meritocracy is "a lie. Merit remains what it has always been: a myth. Pedigree, race, class, gender, politics or something as simple as good looks might not determine sports outcomes as it might, say, Ivy League admissions, but it has always affected the final score—especially if you happen to monitor more than just points."[11] What makes technoscience so troubling is that it threatens this meritocratic ideal. Fame-altering technoscience disrupts the construct of meritocracy. Sadly, a pure meritocratic sporting world probably never existed, but since the mid-1960s the pace at which technoscience has impacted sport and the degree to which it became more visible increased substantially. Herein lies the perception problem with technoscience in sport.

In the span of a few decades, many sports quickly migrated from century-old equipment and "natural" materials such as wood and wool to new laboratory-created substances such as carbon fiber and spandex. Sports such as golf and hockey began the process of abandoning wooden club shafts and sticks in the early 1970s, and by the turn of the twenty-first century, wood had virtually disappeared. Similarly, wool, known for its wicking properties, had been the chosen athletic clothing material for most of the twentieth century, but newly engineered fabrics replaced wool for a host of reasons such as flexibility, lightness, and breathability. The seemingly quick transitions across all sports heightened a larger sporting public's understanding of the use of technoscience. But in certain areas, the new advanced technoscience became a problem when it was seen as playing an increasingly important role in the game and undermining sporting meritocracy.

Framing Technoscience

The idea that technoscience is part of sport is familiar and not particularly controversial. Even for the most casual viewer, it is understood that someone applied scientific and technological knowledge to create the equipment that athletes use. But the place of science and technology in sport has profoundly changed over the past several decades. For instance, recent Olympic Games have been showcases of technoscientific ingenuity. During the 2008 Beijing Olympics, broadcasters reveled in comparing Michael Phelps's eight gold medal performance with Mark Spitz's seven gold medals received in 1972. The comparison of Olympic medals probably intrigued most viewers, but for me the nearly unacknowledged visuals were much more interesting. The mustached, sagging stars-and-bars-suited Spitz with a full head of hair contrasted starkly to the smooth-shaven, hydrodynamic, full-bodysuit-wearing Phelps. Multiple media outlets picked up on these differences but did not push further to examine what they meant for swimming or the larger world of sport.

To more fully understand the evolving relationships between technoscience and sport, studies need to move beyond basic observations of historical variation to inquire what is at stake for the future evolution of sport in an age where scientific knowledge and technological artifacts outstrip original aims of sporting competitions. To acquire this holistic viewpoint, it is necessary to understand what it means for publics as consumers, athletes as users, engineers and scientists as producers, and governing bodies as regulators to see, experience, create, and legislate technoscience's growing omnipresent reach into sport.

Drawing from actor network theorists such as Bruno Latour, it is imperative to think about sport as a community of human and nonhuman actors to more fully examine how technoscience functions as a relational form of life among and between different segments of sporting cultures.[12] The present volume focuses on publics, governing bodies, competitors, and technoscientific actors as the four main groups constituting sporting communities. Publics include those who consume athletic competitions. They can be fans, detractors, journalists, media critics, or casual viewers. These groups often respond most vocally when technoscience seems to sublimate the physical body to a device and give one competitor an "unfair" advantage. Conversely, very little apprehension is seen when governing institutions implement technoscientific monitoring devices or systems, such as instant replay, to increase refereeing accuracy as a means to support a version of "fair" play.

Sporting publics have a great deal to say about what, where, how, and when technoscience influences their beloved athletic competitions. Though publics can be extremely vocal, sport governing bodies, in their charge to create, manage, and defend a whole spectrum of sporting practices, ultimately determine what is and is not permissible within sporting competitions. Since sport governing bodies define the parameters of competitive play, they possess the power and authority to legislate technoscientific use. This legislation ranges from basic specifications, such as the legal dimensions and air pressure of a football, to outright banning certain technoscience, such as in the case of a "corked" baseball bat. Governing bodies and publics have a symbiotic relationship. Governing bodies need the publics to consume competitions just as much as the publics need to trust that the governing bodies will continue to maintain the elevated position of the athletic body within sporting competitions.

Competitors and technoscientific actors are less critically engaged in commenting on the growing role of technoscience in sport. Competitors receive validation primarily through winning. The rewards structures of elite-level and professional sports demand athletic excellence. Where small variations can mean the difference between winning and losing, elite-level athletes ferret out every possible option to gain a competitive advantage. In this regard, it has never been in any competitor's best interest to have a level playing field. Though one can argue that sport is the last bastion of meritocracy, competitive sport necessitates gaining and exploiting inequalities. Mismatches between a taller player and a shorter player, a heavier player and a lighter player, or a faster player and a slower player can produce highlight-worthy plays. Fans see these mismatches as just part of the game, and great players and teams take advantage of these natural discrepancies. In the human drama of sport, fans revere athletes and celebrate the plays where a competitor overcomes her genetic limitations and competes well against a seemingly superior athlete. Doing the seemingly impossible makes sport exciting and draws the public into the narratives of sport. Yet fans and governing bodies do not hold the same esteem for technoscientific mismatches as they do for bodily incongruities.

Publics often equate technoscience that enhances an athlete's performance with cheating. The intellectual creativity exhibited by engineers and designers can transgress the boundary between natural and artificial. These are the sacred yet ever-moving lines that determine what is and is not acceptable. Increasingly, sport's financial, social, and cultural reward systems propel athletes to seek these advantages, the benefits of which have become increasingly more visible. In sporting worlds, where the differences between finely tuned athletic bodies are very slight, athletes gravitate to new and emerging technoscience—hopefully to which their competitors do not have access—to gain an upper hand. Athletes have symbiotic relationships with the technoscientific actors who develop the devices that can, in the most ideal situations, give an athlete a better chance of winning. In this role, technoscientific actors design, develop, and create game-changing artifacts or systems. Technoscientific actors produce devices that may give athletes the biggest competitive advantage while simultaneously creating value for a consumer product brand. However, these devices must not run afoul of the public's desire to believe that bodies are solely responsible for winning sporting competitions or a sport governing body's need to maintain the illusion of balanced and technoscientifically limited playing conditions. Major sportswear brands, such as Nike, Adidas, and Puma, are extremely effective in walking this sociopolitical tightrope.[13] However, smaller companies that manufacture specialized products can struggle to stay out of the crosshairs of governing institutions' regulatory aim.

These four constituencies—publics, governing bodies, competitors, and technoscientific actors—maintain an uneasy equilibrium regarding technoscience. Historically, they preserved this equilibrium fairly easily, but the efficacy of technoscientific augmentation increasingly tests the public's trust and governing bodies' ruling authority. This reality places multiple sporting cultures at the nexus of a new set of tensions between man and machine. Currently, there exist three primary options for any new competition-changing technoscience: acceptance, banishment, and denial. Denial tends to create more problems, so endorsing or prohibiting technoscience are the only two reasonable options. Sadly, technoscience in sport is increasingly becoming about the push and pull between the "authentic" athletic body and "artificial" machinelike technoscience. The relevance of technoscience in sport will only increase, and the ways in which sporting cultures incorporate or suppress technoscience will define the future of athletic competition in the current century.

Framing the Athlete

How does one define a pure athletic performance? The idea that one could discern, measure, and verify an authentic athletic performance is highly problematic. Yet sport governing bodies work very hard to institutionalize processes and protocols to convince the public, as well as athletes, that though they may not reach the point of unequivocally adjudicating authentic athletic performance, these organizations can legislate systems of rules—undergirded with technoscientific tests—that will produce, to the best of their ability, fair competition. The idea of testing the body's performance and using mechanistic metaphors to describe its action has a long history.[14] In the first few sentences of the introduction to Thomas Hobbes's *Leviathan*, he wrote: "For what is the heart, but a spring; and the nerves, but so many strings; and the joints, but so many wheels, giving motion to the whole body."[15] Mechanical views of the body, as illustrated by Hobbes, encourage mechanistic interpretations of the body. The body can become a test site, and the compression ratios of the body's "springs" can be measured as effectively as genetic material can be tested and evaluated. The process of quantification to make evaluative assessments about bodies can be a useful way to understand basic physiological differences, but the utilization of technoscience for bodily assessment has an equally dark history.

Science and technology have been used as fulcrums to boost deleterious racial, social, cultural, gendered, and genetic agendas.[16] Donna Haraway, Katherine Hayles, and many others have written thoughtful critiques of what it means to struggle with conceptualizing the body as a machine as well as its integration with machines.[17] The framing of the body as a machine has profound implications for the future of sport when efforts are being made to create a larger gap between athletic bodies and the devices that competitors use. By subjecting athletic bodies to a host of technoscientific tests for admission to competition, sport governing institutions may not be that far from the technoscientific actors that might view athletes as running and walking laboratories.

What does it mean for athletes, fans, and governing organizations to conceptualize, understand, and accept the human body as an analyzable and endlessly improvable piece of equipment? Jan Rintala argues that this progressive vision dehumanizes and alienates athletes from their bodies and their chosen sporting community.[18] This analysis makes sense in a worldview embedded in historical notions of gentlemanly athletic competitions of the early twentieth century. But this perspective becomes increasingly less relevant in sporting cultures that dismiss the dynamics of dehumanization and alienation because technoscientific interventions are necessary to keep sport safe, rebuild injured bodies, maintain an upward slope of human performance, make the game more entertaining, and sustain public trust. Not surprisingly, these agendas fit lockstep with the Olympic motto of "Citius, Altius, Fortius," or "Faster, Higher, Stronger." As the drive for bigger, stronger, and faster moves forward, there will be more insistent concerns that science and technology have gone too far. In a technoscientific era, athletic competitions may no longer be between humans on the playing field but instead battles between well-funded technoscientific teams. Will athletic competition transcend the athlete and become a competition between the best scientific knowledge and technological methods applied to the human body? Will sport become a medium to display the latest technoscientific innovations?

This book explores these questions, but not in an attempt to argue for a version of technoscientific agency or to examine how technoscience props up or undermines certain sporting communities. It instead is focused on the workings of technoscience within sporting cultures. Specifically, why, within global communities that have embraced both the creative and destructive power of technoscience, do fans, governing bodies, and competitors prefer to push technoscience into an instrumental corner? A partial answer to this question is that revolutionary technoscientific advancements can become unwanted challenges to the tradition, history, and essence of a sporting culture.[19] Clearly, this is reasonable since most sports are activities with deep-rooted and cherished cultural traditions. This is not to say that the participants of sporting communities do not see and understand the power of technoscience, but to maintain the generations-old narrative of sporting competitions and the motor-over-machine parable it must be agreed that technoscience remains neutral, and therefore instrumental. If not, technoscience and the associated network of human and nonhuman actors must be embraced as equal participants within the formation and maintenance of games people play. If technoscience is fully accepted as equivalent to the human participants, these games become less about the athletes and more about the devices, objects, and artifacts in use.

Framing Sporting Cultures

The studies presented here explore how technoscience, in multiple manifestations, participates in both the stabilization and disruption of narrative equilibrium within sporting cultures. One of the most dominant narratives of sport is that all athletic competitions should be fairly decided between competitors on a field of play. This simple premise is deeply loaded. Similar to meritocratic ideals, concepts of what is and is not a fair athletic competition drive this narrative.[20] If nothing changes conceptually, technoscience will continue to develop as a node of power that shapes what sports are and will be. For most of the twentieth century, technoscience maintained a fair balance between competitors. It was part of the regulatory infrastructure, exemplified by agreements on the rules governing categories such as weight classes, event distances, and equipment. More recently, technoscience has become more divisive, raising concerns among athletes, fans, and governing bodies about how it will redefine championed narratives of pure athletic ability.

This analysis specifically uses the term *narrative equilibrium* because it substantively explains the cultural exchanges that take place between the various members of the sporting cultural groups explored. Cultural narratives of sport recount the physical triumphs, mental strategies, and anguishing losses that demand seeing, understanding, and interpreting sport as a human-centered activity. A rich tapestry of cherished historical narratives holds the history of sport together and links celebrated players, transformative games and series, and tradition-laden stadiums to a collective identity fostered by governing bodies, fans, technoscientific actors, and competitors. Each of these groups narrates in its own way. Sport governing bodies narrate through legislation. Fans narrate through beloved stories of triumph and defeat. Athletes narrate through competitive performances. Technoscientific actors narrate through the creation of material artifacts that become part of competitive play. Sporting cultures have equally strong narratives about the authenticity and primacy of the body. Statements such as "she is a natural," "he was born to play," or "she has the gift" firmly reinforce the belief that within sporting competitions, natural bodily talents can separate the best from the rest. Many fields of inquiry, from sport psychology to genetics, study how to extract as much performance as possible from these "natural" athletic bodies.[21] Though equally familiar pronouncements about an athlete's drive, motivation, or work ethic are sometimes deployed to explain why an athlete can be competitive with seemingly inferior physical attributes, it has been difficult to undermine the idea that champions at the elite level are born and not made.

A narrative approach is also a useful way to understand how social and cultural formations bring life and meaning to material objects. In writing about narratives of material culture, Ian Woodward indicates that narratives "come to life by being embodied in objects [that] frequently structure the very way narratives unfold." Furthermore, "objects acquire cultural meaning and power in the context of stories or narratives that locate, value, and render them visible and important. Without such narrative storylines . . . an object is rendered virtually invisible within a culture."[22] This reading of material objects can be effectively applied to technoscience and sporting cultures to examine and explore the ways in which publics, competitors, sport governing bodies, and technoscientific actors narrate the meanings of technoscience into and out of competitive fields of play.

An important aspect of the narration of sport is the politics, and the multiple dynamics of technoscientific "fixing." Sporting cultures can deploy technoscience similar to the way Lisa Rosner writes about the ways in which individuals and institutions unsuccessfully use technology to resolve convoluted social and cultural problems.[23] Sport governing bodies have used technoscience to fix imbalances in sport, as in the cases of improved timing devices, instant replay, and ball tracking, or just to make sure that all players use equivalent equipment. Technoscientific fixing in a sporting context has also been about fixing meaning. Ian Woodward contends that the desire to narrate and fix objects is prevalent "in settings and spaces where the meaning of objects is open to interpretation or debate, or where people are anxious about . . . the meaning of objects."[24] This tension over the meaning of technoscientific objects is so powerful because it can influence a sport's social and cultural identity.

Knowing, understanding, controlling, or fixing the meaning of technoscience relevant to a game is deeply important because the specific use or nonuse of technoscience helps to shape and define a specific sport's culture. The ways in which sporting cultures narrate technoscience is important because these narrations channel historical and nostalgic forces through the circulation and exchange of stories. The power of these narratives is that they inform old and new members of a community about the sporting culture's history, as well as impart the valued perspectives and interpretations regarding the place of technoscience within a specific game. This consistency becomes deeply meaningful for making historical comparisons of records, athletic performances, or teams.

Determining if the 1927 New York Yankees were as good a team as the 1998 Yankees requires standardized metrics. Comparing track and field athletes such as Jesse Owens, Carl Lewis, and Usain Bolt necessitates a belief that each athlete ran under roughly the same conditions. Historically, these sets of culturally sustained collective beliefs avoid fully engaging the changing nature of sporting technoscience because it can potentially destabilize sporting culture's narrative consistency. In the case of Owens, Lewis, and Bolt, the dominant narrative is that their athletic performances can be compared because of the pure athletic simplicity of running. In comparing the 1927 and 1998 Yankees, similar rationalizations are made. The game, at its core, is simple. To compete, all players must master throwing, hitting, running, and catching. The strategies used by players and managers have not appreciably changed over the decades. Combined, they present a strong narrative coherence on which a sporting culture can hang its traditions. Yet it is hard to maintain that the conditions, playing fields, and, most importantly, equipment were the same over time.

If one peels back the outer layers of these narratives, a critical evaluation of sport technoscience can compromise sporting culture's narrative stability.[25] Owens did not have access to the Puma evoSpeed Electric Bolt Tricks track spikes that Bolt wore at the 2016 Olympics or the benefit of competing on the speedy Mondotrack WS synthetic track on which Bolt won his third consecutive round of gold medals in the 100 m, 200 m, and 4×100 m races.[26] Furthermore, Owens did not have access to decades of scientifically proven training methods to prepare for competition. The impacts of these elements, unlike 100 m times, are not easily quantifiable. Moreover, track and field has invested in timing as a constitutive element of comparative equality. Thus to compare athletes over time, game-altering technoscience must be eliminated from the narrative because it precipitates more questions than it answers and may potentially create narrative indeterminacy, where no athlete or team can be compared over time because the conditions of the games were so incommensurable.

Sporting cultures respond to technoscience in varying but deliberate ways. Regularly, the response is directly related to technoscience's relationship with the body and its perceived impact on the authenticity of a sporting competition. When technoscience, which is initially accepted as inconsequential to a specific sport, is seen later as altering the competition, the common response has been to ban the technoscience instead of examining and reevaluating the terms, rules, belief systems, or traditions of a game. But of course, banning any technoscientific product is never simple or straightforward. In her important work on sport, body, and technology, Tara Magdalinski argues: "Although some technologies find a comfortable place in sport, those that are categorically rejected as inappropriately intrusive include any that threaten to fundamentally alter the body and its capacity."[27] Magdalinski's observations and analyses are spot on, but it is important to move beyond merely examining technoscience and its relative distance from the body to understanding and reading the proximity of technoscience to the body against perceptions of authenticity.

Rather than investigating the tensions around the fear of the technoscientific altering of bodies—as in the case of performance-enhancing drugs—these studies examine the social and cultural processes by which technoscience, revealed as more than instrumental, moves from exciting and new to problematic and criticized, and, finally, vilified and banned. This is a crucial transformation in the technoscience and sport dialectic.

In this regard, this book explores the ways in which sport governing bodies, competitors, publics, and technoscientific actors respond to new and emerging technoscience when one or more of these groups no longer views a respective technoscientific object as instrumental.

Interpreting Technoscience: Judging Artifacts and Evaluating Bodies

To gain a fuller understanding of the multiple ways publics, sport governing bodies, technoscientific actors, and athletes negotiate the evolving relationships between sport and technoscience, this book is divided into two three-chapter parts: "Judging Artifacts" and "Evaluating Bodies." Each section explores the evolutionary ways in which different constituencies within sporting cultures negotiate their relationships with the technoscience living within sporting competitions. The interactions described in part I investigate the social and cultural processes by which technoscience reaches a point deemed threatening enough that it must be removed from a sport. Moreover, these sections ask, does banning technoscience really solve the larger problem or does it motivate new technoscientific innovation?

Part II explores how testing techniques and regimes can support or undermine cherished narratives of the sporting body. Specifically, the studies discussed in this section of the book examine how technoscience is deployed to conclusively determine what bodies are and are not permitted to compete. The chapters in both parts elucidate how technoscience unhinges historically valorized sporting narratives. Part I explores swimwear, athletic equipment, and prosthetic limbs to elucidate the ways in which sporting cultures have managed and negotiated new and emerging technoscience. Part II examines gender verification testing, the demise of direct drug testing, and the rise of indirect drug testing with the athlete biological passport (ABP) to explain the problems with technoscientific testing as a determiner of whether an athlete should or should not be allowed to participate in a given sporting competition. Though no discussion of technoscience in sport can be exhaustive, this book aims to present a coherent snapshot of the evolving place of technoscience in sport.

Judging Artifacts

For artifacts of sport, one must begin with the premise that most sports are played with some type of equipment. Though the form, materials, and style of equipment changes over time, these objects have generally been seen as inconsequential. Outside of color, style, or branding, these artifacts typically do not rise to a level of interest or concern. However, equipment has become a problem when it is perceived to give an athlete a significant competitive advantage. The past few decades of elite-level swimming have been defined by the evolution of swimsuit technology.[28] When competitors first began wearing full-body compression fast suits in the late 1990s, the rhetoric about these suits was that they were "faster." Nevertheless, there was not much science outside of proprietary research performed by racing swimsuit manufacturers. Early scientific studies did not fully support the idea that these new suits were markedly faster because a host of issues, from body shape and size to a swimmers stroke, ultimately influenced speed. Though the suits made swimmers more streamlined, they also supplied motivation for swimmers to work harder because they felt that the suits gave them an appreciable competitive advantage.[29] Consequently, confidence, rather than technoscientific innovation, potentially could explain the initial boosts in performances.

The real potent change in swimsuit technology arrived around 2008 with the introduction of Speedo's LZR Racer suit. What made this suit unique was a new application of polyurethane that allowed the suit to reach new heights of hydrodynamic efficiency. By 2009, other companies, such as Arena and its X-Glide suit, pushed polyurethane suits to their technoscientific apex.[30] These suits quickly became a requirement for any athlete who wanted to be competitive. Soon after, cries of "technological doping" began to rain down on the sport, and by 2010 swimming's international governing body, Fédération Internationale de Natation (FINA), banned these types of suits in order to return the sport to the athletes. Yet it remains unclear if banning these devices will curtail technoscience's power on swimming and its associated equipment.[31]

Sports that require some form of running necessitate the use of shoes. Yet athletic shoes traditionally have been seen as important, but not as transformative equipment. Famed stories such as that of Bill Bowerman and the creation of Nike's first waffle-iron-soled running shoe are as heartwarming as they are innovatively genius, but stories of athletic shoes rarely move to the realm of game-changing equipment.[32] However, shoes can profoundly change the outcome of sporting competitions. From Adolf Dassler's handmade shoes for Jesse Owens to wear at the 1936 Olympic games in Berlin to the National Basketball Association's (NBA) ban of shoes by the little-known company Athletic Propulsion Labs, athletic shoes are complex and mildly understood artifacts of sport.

When the NBA introduced a new synthetic ball for the 2006–2007 season, it was the players, not the governing institution, that led the charge for its removal. In an effort to advance the game, the NBA deemed it necessary to use a new, technoscientifically designed basketball. However, the players, who did not participate in the decision to use the new ball, contended that it fundamentally altered the game and made it unsafe. In effect, the players argued that the ball was not an instrumental, mundane, or interchangeable piece of equipment but a central component of how they played game. Their livelihoods depended on the accurate manipulation of this ball, and as a result they demanded a return to a leather ball because it was the ball they had become intimately familiar with. The players, by rejecting the synthetic ball, also inadvertently chose to express themselves as the rightful keepers of the histories and traditions of the game.

In the world of professional cycling, its governing body, the Union Cycliste Internationale (UCI), has made a concerted effort to sublimate bicycles to the body. Eddy Merckx—the undisputed greatest cyclist ever—and his 49.431-kilometer (30.715 miles) hour ride in Mexico City reveals the sporting culture's power to diminish the importance of technoscience. The cycling community believed that Merckx had put the record out of reach and that it would potentially stand in perpetuity. Merckx rode a seemingly standard track bicycle, but when Francesco Moser broke the record twelve years later, his bike was far from ordinary. Moser ushered in the age of aerodynamics for elite-level cycling. From 1984 onward, the bicycles—and the athletes themselves—leveraged every permissible technoscientific aid to push the record farther and farther out of reach. By 1996, Englishman Chris Boardman stretched the record to 56.374 kilometers (35.029 miles) while riding the astoundingly aerodynamic Lotus Type 108 bicycle. In response to this and other technoscientific developments of the 1990s (e.g., performance-enhancing drugs) that questioned the place of the human body within the sport, the UCI deemed it was time to return cycling to its roots and champion human performance over technoscientific innovation.

The UCI decided to rewrite its record books and wipe away all the hour records that came after Merckx.[33] Merckx, his ride, and his bicycle quickly migrated from a historically valuable record-breaking set of artifacts to represent a fundamental shift in the way the sport of professional cycling would use and interpret technoscience and athletes' bodies. The UCI deployed Merckx to curtail the public flogging that the drug-challenged sport received in the press during the last five years of the twentieth century. It comes as no surprise that the UCI, in an effort to maintain its brand, returned to its most valued living legend, Eddy Merckx, and to a perceived moment of pure cycling performance and achievement—when a man was truly mightier than the machine. Yet the irony is that Merckx's machine was far from ordinary. It was the cutting edge of technoscientific design expertly executed by master frame builder Ernesto Colnago. Thus returning a sport to a historically valorized point in time does not mean that it is any purer of a technoscientific moment.

Technoscience can also influence debates and tensions about what it means to be an able-bodied or disabled athlete. Prior to the unraveling of Oscar Pistorius's heroic story with the conviction for the murder of Reeva Steenkamp, the bilateral transtibial amputee's extraordinary sporting records were moving and awe-inspiring.[34] His accomplishments within the Paralympics, however, were understood in a vastly different way in an able-bodied Olympic context. Pistorius's J-shaped carbon fiber prosthetic limbs, designed and manufactured by the Icelandic company Össur, prompted questions about performance enhancement, unfair advantage, and ineligibility. Additionally, and predictably, the world of track and field feared a post-Pistorius, high-tech athlete invasion.[35]

At first glance, it seems obvious that prostheses violate the International Association of Athletics Federations' (IAAF) ruling of March 26, 2007, which prohibits "any technical device that incorporates springs, wheels or any other element that provides the user with an advantage over another athlete not using such a device."[36] In fact, many commentators suspected that the IAAF devised the rule to prevent Pistorius's Olympic bid.[37] The IAAF denied such a motive. Nonetheless, based on testing for an unfair advantage in January 2008, the IAAF determined that Pistorius was ineligible to compete in the Olympics. The IAAF used a December 15, 2007, report by Professor Gert-Peter Brüeggemann of the Institute of Bio-mechanics and Orthopedics at the German Sport University in Cologne to reach its conclusion.[38] The most relevant part of the report states that "in total the double transtibial amputee received significant biomechanical advantages by the prosthetics in comparison to sprinting with natural human legs."[39] Four months later, the Court of Arbitration for Sport (CAS) reversed the IAAF's decision. Redefining what constituted unfair advantage, the CAS declared Pistorius eligible because its own testing showed that the Össur prosthetics offered no net advantage. The IAAF did not contest the CAS ruling and agreed to allow Pistorius to run if he produced an "A" standard qualifying time before the 2008 Olympics, but, unfortunately, he was unable to achieve that mark of entry. He did, however, qualify for the 2012 Olympic Games and became the first athlete to wear prosthetic limbs in the able-bodied Olympics. This reality only fueled the fear—and excitement—about technoscience reconfiguring athletic competition.

The Pistorius case exemplifies how meanings about embodiment are routinely contested and negotiated. It demonstrates how such challenges reinforce—through the concept of fairness—an already-presumed notion of what a sporting body is and should be. Pistorius's eligibility relies on the processes of normalization that locate and classify bodies as acceptable for competition. But we should not lose sight of Pistorius as yet another in a growing number of examples that expose the failure of the modern sporting paradigm. Indeed, as Pistorius and the CAS's ruling that allowed him to compete against able-bodied athletes draws attention to the complexity of science, technology, and sport, they also draw attention to the instability of all bodies and, by extension, the inadequate conceptualization of the athletic body as sport's sacred ground. It seems that sport governing bodies have taken on the responsibility of policing this ground because of their social, cultural, and financial investments in protecting the illusory sanctity of human competition. Ironically, these governing bodies have begun to rely more heavily on the power and authority of science and technology to protect their sports from unwanted, unwarranted, and undermining technoscientific practices and devices.

Evaluating Bodies

Sport is as much about competition as it is about evaluation and testing. During events and training, athletes measure themselves against others. Seeing how one athlete compares to another is a fundamental element that draws publics, governing bodies, technoscientific actors, and athletes to sport. The quest for the most effective evaluative tool, technique, or process supports entire industries. Organizations such as the World Anti-Doping Agency (WADA) and the International Olympic Committee (IOC) have done their best to test athletes within a set of legal, fiscal, geographic, scientific, and technological constraints to ensure that competitions are as fair as possible.[40] But what function does testing serve in sport? There are many answers to this question, but in the sporting context, testing, at least superficially, determines which athletes, or bodies, can and cannot compete or will and will not be sanctioned—all in the name of fairness in sport. Increasingly, governing bodies, publics, and athletes themselves appeal to technoscientific tests to "answer" this question. But should anyone use the results from technoscientific testing to determine who can and cannot compete or who will and will not be sanctioned? Many would argue, absolutely not.

Testing has been deployed to divide sporting competitions by sex. Historically, this has been easy because it is seemingly familiar and simple to understand. When questions arose about sex, it could be tested in the form of gender verification testing. Twenty-first-century variabilities in sex and gender identity, however, make male/female testing seem significantly out of date. Though these tests have been criticized and mildly destabilized, sport governing bodies have not abandoned them completely.[41] What if sex testing within sport performed different work? Instead of using verification testing to distinguish between males and females, what would happen if sporting cultures embraced the continuums of gender and sex to rearticulate these and similar testing instruments to create more interesting and relevant competitions?[42] By winning the 2009 800 m world championship, South Africa's Caster Semenya raised the possibility of this option.

Throughout Semenya's career, her powerful physique has raised questions among her competitors about her sex and gender. Initially, the IAAF responded to these concerns by pulling her from competition in the fall of 2009, only to reinstate her in 2010 with no indication of what decision had been made regarding her sex and gender. Sadly, this situation was not any clearer when she won the 800 m gold medal at the 2016 Rio Olympics. This is unfortunate because in a world where gender, sex, masculinity, and femininity are getting murkier by the day, guidance by one of the world's largest sport governing bodies would have been a practical data point. It also emphatically signifies the limits of these types of evaluation. Despite many tests, the outcome was no clearer than at the start. But perhaps it does not need to be any clearer because it is the binary structure of sex-based competition that should be questioned instead of an athlete's body. After several decades of attempting to understand the place of sex and gender in women's sports, the IOC has given up on gender because sex was seemingly easier to quantify. Science has been so good at quantifying, measuring, and classifying the world, so why wouldn't it work for sex? This decision seemed scientifically plausible because with most everything in sport, someone has attempted to quantify it. It is disappointing that the IOC did not try harder. It seems as if the IOC learned enough to realize that acquiring an incontrovertible understanding of gender is impossible, so it took the easy way out by recusing itself without even attempting to affirmatively state that both sex and gender are social and cultural constructions rather than definitive bodily categories.

When the IOC and various international sporting federations endeavored to contain the explosion of anabolic agents within sport in the late 1960s, they began the process of developing a robust network of governing body interests, athletic concerns, and technoscientific instruments. Though far from perfect, over the next few decades the anti-doping movement effectively enrolled human and nonhuman actors into these anti-doping networks.[43] Currently, these networks struggle to maintain their position of power and privilege because recent events display how ineffective the entire process is for determining wrongdoing. The list of athletes who have admitted to using banned substances but were never caught, as well as the list of athletes whose mishandled samples or incorrect testing protocols enabled them to avoid sanctioning on technicalities, is long.[44] Though the anti-doping movement bolsters its vision from outward attack by raising its collection, transportation, and testing protocols to forensic science levels, it is questionable if making anti-doping more "technoscientific" will assuage larger public concerns about its efficacy.

Testing regimes allow for the abandonment of the important balance between technoscience and culture. Initially, technoscientific testing supported heavily negotiated ruling decisions, but as technoscientific tools migrate from support mechanisms, to enforcement devices, to, finally, rule-making instruments, sport may begin to lose sight of its social and cultural goals in exchange for reining in a runaway drug culture that many governing bodies see as undermining, and eventually destroying, the integrity of their sport. If sport is supposed to be a microcosm of our society, let us bring a bit more of society back in and have more collaborative and collective discussion about the place of technoscience within sport.

The primary model of enforcement and deterrence was and still is direct testing. But over the past decade, strides have been made to create new tools that move from the publicly conclusive measures of direct testing to the potentially gray, fuzzy, and publicly indiscernible black-boxed methods of indirect testing. Thus the anti-doping movement appears to be, at least publicly, moving away from red-handed doping confirmations to multiple experts analyzing data sets that allude to some form of misconduct. In an effort to curtail drug use and maintain the illusions of clean and pure sport, sport governing bodies, fans, and athletes appear to have forgotten that technoscientific testing began to support and not define who can and cannot compete or who will and will not be sanctioned. Some sports simply avoided testing altogether until recently. In the United States, 2012 marked the debut of blood testing for HGH in Major League Baseball (MLB). This plan made MLB the first major sporting league in North America to test its players for HGH. MLB can only be given a partial congratulation, however, because though minor league players are tested all year, those in the major league are only tested in spring training and in the off-season. In 2012, the National Football League (NFL) was much further from testing than MLB. Though the National Football League Players Association (NFLPA) agreed to HGH testing in 2011, it spuriously contested the legitimacy of the test, halting its implementation. There seems to be a valid reason why the NFLPA derailed the testing. Former NFL quarterback Boomer Esiason implied that as much as 60–70 percent of NFL players used HGH.[45] For testing to be effective, it first must be implemented.

The United States Anti-Doping Agency's (USADA) case against Lance Armstrong represents the most recent death of the direct testing model. Armstrong has never been sanctioned for having banned substances in his blood or urine. The evidence wielded against him did not contain any direct testing evidence. Since the USADA was able to take down Lance Armstrong without direct testing, there is conceivably no longer a need for this type of assessment. But will the public accept this change? Anti-doping organizations have so heavily hung their sanctioning authority on the perceived technoscientific power of direct testing that other forms of evidence may not be as convincing. The anti-doping movement persuaded the public to believe that direct testing would provide irrefutable evidence of illegal behavior, but this push did allow the anti-doping movement an exit plan if and when the ineffectiveness of direct testing was exposed.

If Lance Armstrong represents the death of testing, then the ABP is the resurrection. In the early 1990s, EPO became the most potent drug for endurance sport. Its effectiveness in increasing the oxygen-carrying capacity of blood, to unhealthy levels, was unquestionable. The ABP is a response to this and other similar drugs. In 2006, the scientific, technological, and cultural impetuses coalesced in the WADA's Haematological Working Group. This group supported the broad idea of an "Athlete's Haematological Passport" and chose a model devised by a group of researchers led by Pierre-Edouard Sottas.[46] Sottas's group made a case for using statistical classification techniques to develop a more comprehensive method of determining if athletes used blood-boosting drugs such as EPO or increased their performances with blood transfusions. Early on, there was great enthusiasm about the ABP significantly curtailing all forms of doping. The UCI emphasized that new, indirect modes of detection would bring a higher-level technoscientific rigor to effectively uncover offending behaviors. Though other sport governing bodies, such as the IAAF, are buying into this tool, scientists are less convinced of its regulatory power.[47] In fact, very recently international journalists have exposed a problem with this model by showing how a technique called microdosing can effectively subvert the ABP's testing protocols.[48]

Just as sport is a defining element of the modern world, technoscience's material and conceptual power influences how societies define themselves and fulfill hopes and dreams about brighter futures. Since the mid-twentieth century, sport and technoscience have become two of the most dominant and defining cultural and societal narratives. Technoscience has been and will always be part of sport, though some may not want to embrace its power. This book is about understanding the power of technoscience and what that power means for the future of sport. Collectively, these case studies illustrate that in certain instances technoscience must be disciplined and eventually excised from sporting cultures to secure a certain future. What that future of sport becomes is difficult to determine, but until a new understanding of technoscience is embraced, the future of sport will be steeped in traditions, histories, and nostalgias that will continue to battle against technoscience in order to maintain the primacy of the human body in all athletic competitions. The space between sporting traditions, bodily ability, and random acts of life is the messy place where technoscience resides within sport. It delicately dances between athletes' needs, publics' desires, engineering visions, and governing bodies' legislation. The subsequent chapters aim to begin the process of understanding these ever-evolving, culturally embedded technoscientific relationships.

I JUDGING ARTIFACTS

1

Black Is the New Fast

Swimsuit Technoscience and the

Recalibration of Elite Swimming

An impassioned Bob Bowman exclaimed at the 2009 Fédération Internationale de Natation World Championships: "We've lost all the history of the sport . . . the sport is in shambles right now and they better do something or they're going to lose their guy who fills these seats." His exclamations came after years of controversy around the design of swimsuits and reconfirmed that swimming had become enveloped in an absurd and surreal technoscientific world. That "guy who fills these seats" was Michael Phelps, Bowman's primary pupil. He couched his plea for the future of the sport in a complex narrative concerned with Phelps's waning performances to those wearing "faster" fast suits and the marketing of competitive swimming when its most marketed swimming hero was no longer winning and potentially on the verge of becoming uncompetitive and irrelevant. His carefully worded statement concluded by proposing a compromise. He suggested that Fédération Internationale de Natation "adjust all the records starting with the LZR [Speedo's record-breaking swimsuit]. If we took them all out and went back to 2007 . . . even [the records] in Beijing. We can have them in a separate list. These were done in polyurethane suits and then these are done in textile suits. Then we can start over in January and make the sport about swimming."[1]

By the time Bowman uttered these concerns, swimming had already materially and metaphorically dived off the deep end into a world where swimsuit drag coefficients were vastly more important than the number of training laps any athlete had done in the pool. Bowman's suggestions included turning back the clock, splitting record books, and changing the way FINA handled its business, but by most accounts, it was unclear where and how to begin without turning the sport upside down and dismantling the last elements of its legitimacy. Bowman's comments at the 2009 World Championships marked the end of a cycle that forever changed the record books of swimming but, more importantly, demanded that publics, athletes, and governing institutions no longer ignore the potent effects of technoscience on competitive swimming.

Over the past few decades, elite-level swimming has been featured prominently in public discussions regarding technoscience and its impact on athletics. Athletes clad from ankle to neck in the black technoscientifically engineered materials became de rigueur for any swimmer wanting to compete for victory. Swimsuits migrated from benign tools crafted to cover the body and display an athlete's country of origin to an obligatory technoscientific device necessary to be competitive. Swimming analyses quickly turned away from biometrics such as height, stroke rate, arm length, and foot size to obsessive examinations of who wore which company's newest suit. The technoscientific power of fast suit technoscience completely subsumed narratives of athletic ability. This technoscientific narrative became so powerful that FINA chose to push the reset button and return swimming to an earlier period by banning the types of suits that had produced "unnatural" explosions of performance that quickly rewrote swimming record books in less than a decade. These swimsuits undermined beloved narratives of human athletic achievement and as a result had to be removed from the sport to maintain the cherished primacy of the human body.

Though late nineteenth- and early twentieth-century suits were made of natural fibers such as cotton and wool (instead of modern lab-based materials such as Lycra and polyurethane), athletes have always chosen to wear the fastest and most competitive suits. Images of elite swimming over the twentieth century narrate a history of smaller and tighter-fitting suits. Though this story is not exempt from questions about sexual mores, it is more powerfully a technoscientific timeline of the increasing importance of swimming and winning, and how bodily hydrodynamic efficiency is paramount to swimming success. This is why the introduction of ultrafast technoscientific swimsuits should not come as a surprise. It should, rather, be viewed as a logical conclusion to a century-long effort to swim faster. Swimming fast will always be a battle between athletic ability and hydrodynamic drag. Though early twentieth-century swimmers did not use such specific scientific terminology to describe the natural affordance of water, reducing the drag of an athlete in the water directly correlated to faster propulsion. Currently, at elite levels the differences in athletic ability between swimmers tend to be small, and the reduction in drag has proven, in many instances, to be the difference between winning and losing. Bob Bowman's comments at the 2009 FINA World Championships displayed the frustration felt by many that modern technoscientific swimsuits had become vastly more important than the athletes. So how did swimming get there?

The Modern Olympics and Swimsuit Design

Competitive swimming has a long history. It was one of the events athletes contested in the reimagined modern Olympic Games in 1896. Swimmers only raced four events in this first Olympics, the 100 m, 500 m, 1200 m, and a special 100 m race for sailors in the Greek Royal Navy. Eighteen-year-old Hungarian Alfréd Hajós won the 100 m and 1200 m events. He wore a form-fitting, mid-thigh-to-mid-bicep suit that could easily be interpreted as a precursor to the suits of the past few decades. Women first competed at the 1912 Olympic Games. They competed in only two short events, the 100 m and 4×100 m relay, because the International Olympic Committee questioned if swimming any farther unduly taxed a woman's body. England's Isabella "Belle" Moore and her 4×100 m teammates Jennie Fletcher, Annie Speirs, and Irene Steer (figure 1.1) became the women's swimming stars of these games[2] and wore suits similar to that of Hajós. These British women not only pushed societal notions of female propriety by wearing suits exposing large expanses of skin but also challenged cultural notions of femininity by merely competing. The resolute woman standing in the middle covered from neck to foot in a dark dress is in stark juxtaposition to the swimmers and their athletic garb.

Understanding the conservative social conventions of dress before the 1920s, it is somewhat surprising to see female Olympic swimmers wearing form-fitting suits. Yet, at another level, it is not really that surprising. Hajós as well as Moore and her teammates attended the Olympic

Figure 1.1. England's gold medal winning 4×100 m team of Isabella Moore, Jennie Fletcher, Annie Speirs, and Irene Steer at the 1912 Olympic Games. The fact that they wore the most advanced silk suits weighing in at about two ounces illustrates that technoscience has always been a key component of swimming. George Grantham Bain Collection, Library of Congress

Games with the intention of winning medals. All were knowledgeable swimmers and undoubtedly knew that heavier and looser-fitting suits sapped medal-winning speed. Thus, contrary to recent hand-wringing around the technoscientific suits of the first decades of the twentieth century, the efficiency of suits and bodies has always been part of the sport and business of competitive swimming. Some of the first acknowledgments of the importance of swimsuits can be traced to the early twentieth century and a company founded in Sydney, Australia, in 1914.

Alexander MacRae founded MacRae and Company Hosiery, soon to become MacRae Knitting Mills Ltd., to manufacturer wool and cotton undergarments. The company chose the formidable name of "Fortitude" for its underwear. In the early 1920s, MacRae first dabbled in the swimsuit business to supply the substantial market of beachgoers. The company also began building relationships with swimming associations in Australia and, eventually, abroad. The suit they designed had a "racer-back" style with large arm openings and narrow shoulder straps similar to many suits of the day, but the innovation resided in the manner in which the straps connected between the athlete's shoulder blades. This design allowed for a more form-fitting suit that conceivably produced less drag than the standard tank top suits of the era. By the late 1920s, the company began to make significant investments in swimwear manufacturing as well as supplying suits to elite swimmers.

Multiple-time European champion Claes "Arne" Borg from Sweden became the 1500 m Olympic champion in a MacRae suit in 1928, and Australian swimming phenomenon Andrew "Boy" Charlton is also known to have worn MacRae suits. MacRae himself ran an internal competition to create a catchy name and slogan to sell his company's gold medal winning suits. Retired sea captain Jim Parson won the competition and five Australian pounds for "Speed on in your Speedo." MacRae confirmed his company's strong presence in swimsuit apparel in 1929, when he changed its name to Speedo Knitting Mills.[3]

Borg, Charlton, and American Johnny Weissmuller dominated men's swimming in the 1920s and exchanged world records multiple times. Most images from competitions in the 1920s show them, as well as their female counterparts, wearing suits of a similar design. By the early 1930s, most, if not all, competitive swimmers were swimming in some version of the racer-back suit. These suits drew little concern regarding performance enhancement, though Australian breaststroker Clara "Clare" Dennis was temporarily disqualified at the 1932 Los Angeles Olympic Games because of a concern that her suit did not cover enough of her shoulder blades.

By the mid-1950s, the lightness and strength of silk supplanted wool and cotton as the chosen material. Speedo began building suits in the slicker, smoother, and more hydrodynamic material of nylon. Besides strength and elasticity, nylon also did not absorb water like natural fibers. Therefore, nylon suits overcame some of the major hydrophilic limitations and the expense of historically preferred materials such as silk. Speedo continued to push the cultural, social, and technoscientific boundaries within swimming for the rest of the century, to the degree that Speedo has become a euphemism for a small and revealing swimsuit.

Over the twentieth century, leaps in swimming performance can be attributed to swimsuits, but the improvements in the swimming surface, swimming technique, and athletic preparation all enabled athletes to swim faster. In 1896, Hajós won his gold medals in the Mediterranean Sea in cold, windy, and wavy conditions in which the water temperature hovered around the midfifties and waves crested ten feet. This is a far cry from contemporary competition pools, housed in climate-controlled structures containing finned lane lines squelching up to 70 percent of wave and wake energy with water temperatures falling between the FINA-prescribed parameters of 77 and 82 degrees. In 1896, athletes could employ any stroke to propel themselves through the water. Hajós used the "*sailor tempo*, which featured the paddle-like underwater movement of the palms."[4] This dog-paddle style gave way to the freestyle stroke, and the Olympics proliferated strokes that now include the backstroke, butterfly, and breaststroke. Technique has and always will be a central focus of elite swimming, and a great deal of contemporary research concentrates on ways to create data analyses of stroke efficiency and propulsion.[5]

The preparation of athletes in the early twentieth century has very little resemblance to training today. Contemporary physical and mental preparation of swimmers benefits from a century of work on training across sport, and improved performances attest to its success.[6] For the most part, these technoscientific advances did not raise significant concern because they did not supersede the athletic body, and in cases where the competitive environment changed, such as pool design, it appeared to impact all athletes equally.

From the 1950s onward, companies such as Speedo began to introduce more technoscience into the world of swimming to decrease drag and allow athletes to swim faster. By the 1972 Olympic Games in Munich, Speedo had developed competition suits using spandex. This synthetic fiber patented by Joseph Clois Shivers Jr. and assigned to DuPont in 1962 became the basis for the next evolution in competitive swimwear and fashion.[7] In the late 1950s, chemists at manufacturers such as DuPont attempted to synthesize fabrics possessing more stretch and elastic recovery than natural materials such as rubber. The proliferation of synthetic fibers revolutionized the textile industry, but early materials had limitations. They did not stretch or return to their original shape particularly well. Rubber-based products had been used in these types of applications, but rubber's long-term performance and durability limited its effectiveness. Spandex, more popularly known by DuPont's trade name of Lycra, substantially filled this material void.[8] Designing athletic apparel, especially competition swimsuits, provided an outstanding opportunity to display the material's elastic properties. The ability to create nearly form-fitting swimwear helped to overcome elite swimming's nemesis of drag. The results of the swimming events of the 1972 Olympic Games confirmed the material's effectiveness. Speedo dominated the elite swimming market to such a degree that athletes wearing Speedo suits broke twenty-one of the twenty-two world records in Munich.[9] This outcome, along with the emergence of new material science, intensified the focus of swimming on drag and the dynamics of swimwear.

By the mid-1980s, repelling water garnered much research attention, and scientists began to focus on "minimizing the area of contact with the water and on materials that would expel . . . water that entered the suits."[10] This design philosophy informed the technoscience beneath racing swimsuits until roughly the 1992 Barcelona Summer Olympic Games, at which Speedo became the first to bring a "fast suit," the S2000, to market. Speedo claimed the S2000's material produced 15 percent less drag than prior swimsuit fabrics. The suit was worn by Speedo-sponsored athletes as well as those under contract with Mizuno because of a cross-licensing agreement with the two companies that had begun in 1966. These athletes won more than half of the swimming medals awarded in Barcelona. The S2000 was not only fast, but with its eye-catching, shiny appearance it looked the part as well. The suit also brought a new level of interest— sporting and sexual—to competitive swimming. Athletics has always contained a level of sexual voyeurism, and the suit embraced that aspect of sport.[11] These suits caused quite a stir, not only because of their success in the pool but also because of the sexualized overtures the skintight and rubbery material conjured.[12] Though the suit was visually striking in this context, it is meaningful because it launched contemporary technoscientific warfare within competition swimsuit design.

Engineering the "Fast" Suit

After the Barcelona Olympic Games, a conceptual shift occurred in the way scientists thought about bodies and drag in relation to swimsuit design. This shift was partially driven by the application of technoscientific data analysis methods such as computational fluid dynamics and boundary layer control. Prior to computer-driven analyses of swimming hydrodynamics, most athletes and trainers believed that a smoother stroke and a hydrodynamically "slipperier" body and suit enabled an athlete to swim faster and more efficiently. Athletes maintain speed by gliding through the water rather than paddling. Computational tools allowed researchers to measure bodily dynamics within a swimming pool and displayed how different body shapes and sizes produced varying amounts of drag.[13] Researchers already understood how human flesh became unruly once buoyed by water, but data analysis showed how profoundly problematic a human body can be. To deal with and attempt to control the hydrodynamic issues bodies presented, researchers began to design suits with the goal of turning swimmers' bodies into more rigid, regularized structures of hydrodynamic efficiency. If Lycra had ushered in the era of smaller and tighter-fitting swimsuits, newer designs aimed to increase the importance of the suit by diminishing the hydrodynamic randomness of bodies.

Studies of bodily efficiency in water had traditionally been performed by dragging mannequins through the water using a pulley system. The uniform and unwavering surfaces of mannequins provided an ideal laboratory environment to test suit materials, but they arguably also highlighted the suit's fickleness in relation to real human bodies, as an athlete's body and swim stroke do not correspond to the shape of a mannequin.[14] Even hair disrupts the movement of water across a body, but not in a controlled way. Thus a specific swimsuit's performance is only a relative measurement that may not produce the same level of efficiency for all body shapes. What that means is that to minimize drag, the body's surface has to be altered and controlled. To combat drag and the inefficiencies of human skin and flesh, the body itself became a problem that needed to be fixed. The solution was to encase the body and provide a system by which engineers, designers, and scientists could control the flow of water across the surface of a swimming body. In the late 1980s, promising early research at Tsukuba University showed that dimpled materials could reduce drag. Though the experiments with dimpled suits were not entirely successful, the insights gleaned from this research turned swimsuit design in a new direction.[15]

Boundary layer research would be put to good use in designing the suits for the 1996 Atlanta Olympic Games when Speedo and Mizuno collaborated on the next fast suit: the Aquablade. The technoscientific innovations in this suit were seen in the contrasting shiny and dull vertical stripes. The striped texture, smooth for the shiny and rough for the dull, created alternating fast- and slow-moving flows of water across the surface of the swimsuit. The hydrodynamic vortices that resulted from the differential in current speed forced the water to hug the suit's surface more closely at a higher rate of flow.[16] Japanese studies on these suits determined that using this material for women's suits produced a net advantage of 9 percent over traditional swimsuit materials.[17] To the naked eye, these swimsuits resembled a "traditional" suit. Those unfamiliar with the science behind the suits' design would not know the advantage the suits could provide. These types of scientific investigations led to efforts geared at gaining more control of the body-swimsuit-water interface. In the progressive quest to control the flow of water over skin, the next logical step in suit design was to more effectively encase the body in a hydrodynamic second skin. Speedo had been a leader in swimsuit technoscientific innovation for most of the twentieth century, but Adidas and the Australian swimmer Ian Thorpe paved the way to the popular marketing and consumption of the full-body suit design in the late 1990s.

Ian Thorpe was *the* swimmer of the late 1990s.[18] In 1997, the photogenic swimming phenomenon was the youngest ever, at the age of fourteen, to be selected for the Australian national swimming team. The country had great expectations for him, and he did not disappoint. By winning the 400 m freestyle at the FINA World Championships in 1998, at the age of fifteen, he became the youngest male individual world champion. The championships, held in Perth, Australia, launched his career as a national hero and heavy Olympic favorite. The fervor around him only continued to rise when *Swimming World* selected him as the youngest man to win World Swimmer of the Year in 1999. After a dominant and multiple-world-record-breaking performance at the Pan Pacific Championships in August 1999, held at the same venue that hosted the Sydney Olympics swimming events, Thorpe and Adidas announced that they had agreed that Thorpe would swim in Adidas equipment going forward.[19] The Australian public was shocked that the greatest swimming hero of that moment, and potentially ever, would leave the home brand of Speedo (though owned by the British-based Pentland Group since 1990) and defect to Germany and Adidas.

This was a shrewd move by both parties. Adidas introduced their full-body suit, the Jetconcept, in 1998 and needed an elite athlete to validate its research and development at the upcoming Olympics. A multiple medal winner, Thorpe was a solid bet, and he already possessed a robust popular media following. Signing him as a showcase athlete guaranteed that the Adidas brand and their products reached multiple publics over the next few years. For Thorpe, the move, outside of the financial benefits, placed him in a strong technoscientific position for the upcoming games. Adidas made the conscious choice to focus on the drag created by a swimmer's body shape rather than the frictional drag of the water. They contended that the form drag, or the shape of the body, accounted for seven times more drag than frictional drag, or the resistance produced by a body moving through water. They designed a series of "small 'riblets', integrated into the new suit, [to] channel the water and thereby shift the turbulence that occurs on the swimmer's body and reduce the amount of water a swimmer carries on his back."[20]

Speedo and Arena, like Adidas, also honed in on controlling the form of the body. These companies' design teams and partners innovatively applied the aeronautical engineering principle of boundary layer control to the design of their suits. By inverting the belief that absolute smoothness was the key to swimming speed and efficiency, boundary layer control supported a conceptual design shift. It showed that uneven surfaces, like the dimpled surface of a golf ball, offered hydrodynamic advantages that reduced the drag of a swimming body. Even though the application was not exactly straightforward, this revelation drove the designs of Speedo, Adidas, Arena, and many other companies wanting to compete within racing swimsuit design. Much of the prior boundary layer control research had been performed on fixed or solid surfaces such as cylinders, airplane wings, and boat hulls, making highly flexible bodies an entirely different matter.[21] The general idea is that if a swimmer's back can be manipulated to perform similarly to an airplane wing, the swimmer can create lift, swim "higher" in the water, and propel his or her body quicker. The initial research findings had to balance rigidity and control against flexibility and comfort. Suits could be built to create a great deal of lift, but it was questionable if an athlete could effectively swim in them. New materials had to be developed to compress the body into a desired shape while simultaneously allowing the swimmer to perform the stroking, kicking, and twisting necessary in competition. To manage these realities, Speedo created three-dimensional body scans of athletes to create custom suits that minimized the movement of bodily flesh and maintained a proper hydrodynamic shape, all the while allowing athletes to swim as powerfully as possible.

The coalescing of new and necessary networks of athletes, industries, and scientific institutions defined swimming in the 1990s, and these relationships manifested themselves in a new range of suits. Speedo developed its own aptly named "Fastskin" suits, which also used a riblet design. Japanese companies Mizuno and Toray Industry Inc., in conjunction with Speedo, developed a material that mimicked sharkskin for their 2000 Sydney Olympic Games suits.[22] Elite American swimmers such as Lenny Krayzelburg, Jenny Thompson, and Amy Van Dyken, who participated in Speedo's testing, spoke proudly about how they could "feel the difference" wearing the Fastskin suit.[23] Swimsuit manufacturer Arena began building its network of technoscientific expertise in 1994 and credits the research of Redha Taiar (biomechanical expert), Michel Joseph (textile expert), and Alexander Popov (swimming expert) with providing the foundation for their "Powerskin" technology used in the 2000 Sydney Olympic Games.[24] By the time the Olympics rolled around, Adidas, Arena, and Speedo clearly succeeded in producing "faster" suits. Thorpe took home a trio of gold medals and a pair of silver medals from the 2000 Sydney Olympic Games and fulfilled the suit's technoscientific promise. Not to be outdone, athletes wearing Speedo suits broke thirteen of the fifteen total world records and won 83 percent of the swimming medals.

Questioning Fast Suits

It had become evident—across multiple sporting arenas—that sport would depend more and more on data-driven research and testing, and by the end of the 2000 Sydney Olympic Games there was no doubt about the direction in which swimsuit technology was going and what forces propelled these developments. It was also around this time that questions began to arise regarding the impact these suits were having on the ethics, authenticity, and purity of sport. Though critics initially did not use the term *technological doping* to describe the impact of new swimming equipment, it was clear to the average viewer that this new equipment was shaping the competitive environment in the pool. Before the 2000 Sydney Olympic Games, the idea that publics would be concerned about the technology was not part of the discussion regarding these new full-body swimsuits.

Athletes and companies such as Speedo and Adidas jockeyed for press attention and highlighted how their suits had made an important leap forward in swimming competition. Little did they know that less than a decade later, this energy, impetus, and corporate competition would bring an end to the technoscientific swimsuit arms race launched after FINA's approval of the Speedo Fastskin—and, subsequently, other full-body suits—in November 1999.

Early uses of full-body suits were not without critics. One of the most outspoken was Brent Rushall, a professor of exercise and nutritional sciences at San Diego State. Rushall saw the suits as undermining the authenticity of the sport of swimming. On April 2, 2000, Professor Rushall submitted a document to Richard McLaren, the Court of Arbitration for Sport's (CAS) on-site arbitrator for the 2000 Sydney Olympic Games. His complaint began as follows:

> Swimming across the world is at an important crossroad. The sport
> could be irreparably changed. Now could be the time to either stand up
> for preserving the activity as a pure sport, or let outside commercial
> interests dictate its development largely for a profit motive. . . . Once
> swimming races were decided by the abilities of swimmers alone. Now
> races could be decided by swimming abilities and equipment. The
> coach-swimmer relationship will no longer be the sole determinant of
> competitive success. It is possible that gold medals could go to the
> swimmer with the best performance-enhancing suit rather than the
> best ability and training.[25]

Rushall's comments reflected concerns with how the technoscience of competitive swimwear was not only pulling the sport away from its history and traditions but, most importantly, away from the value of the athletic body. Rushall's main contentions centered on his belief that full-body suits violated FINA's rule SW 10.7. In 2000, FINA rule SW 10.7 stated that "no swimmer shall be permitted to use or wear any device that may aid his speed, buoyancy or endurance during a competition (such as webbed gloves, flippers, fins, etc.)." With all the press releases, advertising, and athlete testimonials showing that the new swimsuits were weapons of speed, it seemed inconceivable that suits such as the Speedo Fastskin were approved in November 1999. One could potentially understand how they passed the requirements for modesty and did not appreciably improve flotation, but the entire purpose of these suits was to significantly increase speed, which by all accounts they did.

Whether the claims were a component of pre-Olympic advertising hyperbole, Rushall rightly argued that Speedo and Adidas could not have it both ways—that their suits enabled athletes to swim faster but did not improve performance. Rushall interpreted the semantic sleight of hand as a way for Adidas and Speedo to gain market share and financially capitalize on record-breaking suits. He perceived FINA to be turning a blind eye to the technoscience in order to sell their sport as sexy, forward thinking, and exciting through multiple world records at the 2000 Sydney Olympic Games. Rushall objected to the collective spin primarily issued by Speedo, Adidas, and FINA. The first bit of spin came when Adidas started calling its full-body suit an "equipment bodysuit" prior to the 2000 Sydney Olympic Games. Rushall was quick to point out that swimming did not have "equipment." He argued that this was a linguistic maneuver to subvert the Australian Olympic Committee's contractual agreement with Speedo for the 2000 Sydney Olympic Games. Traditionally, swimsuits were called a "costume," as in something an athlete performed in. The term *costume* connotes a benign uniform of covering that has no impact on the game. Transitioning a costume into equipment fundamentally altered what a swimsuit is and how athletes can expect it to function during competitions. Using a speed suit seemed monumentally different than a baseball player choosing a specific webbing for his mitt. Rushall argued that FINA either downplayed or ignored the performance capabilities of the suits. Rule SW 10.7 appeared to concentrate on propulsion, and FINA did not see decreasing drag as an active means of increasing propulsion. Besides transgressing this specific rule, Rushall also felt that these new suits, as performance-enhancing tools, and the sporting infrastructure advocating their use had "intruded upon the honorable and traditional concept of competitive swimming as being human ability against human ability."[26] For him, and others, these suits demoted swimmers from the station of noble and heroic athletes to that of lab rats crammed into the newest piece of equipment to prove that technoscience had indeed superseded the body.

Rushall was clearly aware of the fact that full-body suits influenced swimming competitions, but inconclusive results from early scientific studies after the Olympics quelled some of the concerns about the technoscientific advantages these suits provided. Two major studies, one led by Huub

Toussaint of the Institute for Fundamental Clinical Human Movement Sciences at Vrije University and the other by Nat Benjanuvatra of the School of Human Movement and Exercise Science at the University of Western Australia, raised concerns about the suits but did not unequivocally prove the suits were performance-enhancing technoscientific tools. The Benjanuvatra study examined Speedo's claim that the suits reduced hydrodynamic drag while simultaneously not offering any buoyancy advantage. The claim of reducing drag was a bit broad. Swimmers cope with frictional drag (between the water and the surface of the skin or suit), form drag (from the physiological shape of a swimmer's body in relation to the speed at which a swimmer propels herself), and wave drag (from the multiple disturbances created by several athletes swimming in the same body of water at the same time). Therefore, to talk about drag as one solitary or unified entity was a bit spurious. In this regard, the Benjanuvatra study focused on buoyancy, because regardless if a swimmer's physiological makeup allowed her to gracefully glide or powerfully plow through the water, "increased buoyancy enables a swimmer to use the kick more effectively for propulsion because less effort is needed to counteract the torque created when the centres of mass and buoyancy are not aligned as in a less buoyant, mesomorphic body."[27]

Comments verifying the suit's buoyancy directed the Benjanuvatra group to examine if the swimmer's sentiments were actually true. They employed nine swimmers (five men and four women) to test the hydrostatic weight of the Speedo Fastskin suit against a traditional suit by weighing athletes in water and documenting the suit's drag profile while towing swimmers through the water in a pool. The results showed that when fully saturated, "full-length Fastskin suits did not differ in buoyancy from results of a standard swimsuit."[28] Overall, the fabric did not perform like wetsuit materials, which were more capable of trapping air to provide increased lift in the water. The most striking results were found in passive drag testing, where swimmers kept their bodies as still as possible while being towed through the water.

When dragged by a rope with two arms extended, athletes wearing Speedo Fastskin suits reduced hydrodynamic resistance by 5.5 percent at the surface and 10.2 percent at a depth of 0.4 meters. These results showed that the suits could provide a substantial hydrodynamic advantage. Yet when the researchers performed the same drag test and asked the swimmers to perform a flutter kick, the results changed. At the surface of the water, the Fastskin's hydrodynamic efficiency dropped slightly, to 5.3 percent, while the depth number plunged to 4.8 percent. Though the data showed increased hydrodynamic efficiency with the Speedo Fastskin suits, the researchers also noted several caveats.

First, none of the male swimmers had shaved their bodies, which had long been known to decrease drag, and most elite swimmers regularly shaved for competition.[29] Second, and more important, the study did not create a complete drag profile. Though athletes in the study lightly kicked their legs in one series of tests, this test was not close to providing a complete drag profile. Specifically, it did not measure active drag, or the resistance created in the water when an athlete swims at race pace. Active drag depends on varying pool conditions, the body shapes of athletes, and the dynamic ways in which they propel themselves through the water.[30] The multiple variables, specifically the morphology of human bodies, made it very difficult to make strong universal claims about the ways full-body swimsuits enhanced performance.

Huub Toussaint and a group of researchers at the Institute for Fundamental and Clinical Human Movement Sciences attempted to determine the complete advantage a Speedo Fastskin suit provided when an athlete swam at race speeds.[31] This study questioned if Speedo, in its promotional material, chose to highlight the significant passive drag improvements their Fastskin suit provided and obscure the not-so-significant decrease in active drag. In testing six men and seven women at four different swimming velocities, they found "no statistically significant reduction in drag" when wearing a Speedo Fastskin suit.[32] When the data is disaggregated, it becomes even messier. Overall, men had smaller differentials in active drag than women. One reason for this is body fat. Toussaint's study implied that tight-fitting suits "might prevent large oscillating deformations of subcutaneous adipose tissue when swimming at higher speeds."[33] Stated more simply, by confining or disciplining rogue flesh, tight suits reduced the amount of body wiggle while swimming fast. Since women traditionally have more body fat than men, and body fat influences the hydrodynamic efficiency of a swimmer, tight suits worked "better" for women. The study also indicated that as athletes swam faster, the active drag difference between a conventional suit and a Fastskin suit decreased—the implication being that the faster a swimmer swam, the smaller the advantage.

Most importantly, the study showed that the differential in drag varied from athlete to athlete. The implication of this result was that body shape has more influence on the speed at which an athlete can swim than the swimsuit he or she chooses to wear. Additionally, the suit's fit and tightness significantly influenced its effectiveness. The athletes whose suits were constructed using 3-D models of their bodies most likely benefited from the suits more than swimmers using "off-the-shelf" versions. Swimmers can train their bodies to be more aerobically and anaerobically efficient, but it is exceedingly more difficult to reconfigure the physiological shape of one's body. Athletes cannot change their skeletal structure to more effectively conform to an idealized hydrodynamic form. This study revealed that certain bodies are inherently more hydrodynamic than others at different speeds. It also showed that Speedo Fastskin suits, even if they produced an appreciable hydrodynamic advantage, might not be able to overcome the genetic morphological advantages some athletes possessed. As a result, the full-body suits produced by Speedo, Adidas, Arena, or any other company were not necessarily universal keys unlocking phenomenal performance, and, like any new technoscientific tool, they did not work equally for everyone. The suits were the next evolution in suit design, contending with multiple variables including the type of race and distance, body shape and fitness, and fit and design, all of which had varying impacts on a swimmer's ultimate race performance.

When the 2004 Athens Olympics rolled around, the full-body suit design had become the standard equipment for elite swimmers. It was clear to fans, athletes, governing bodies, and equipment designers that "a new swimming body [had been] born."[34] This new merger of athlete and technoscientific device, or a swimming cyborg, became the standard by which all things swimming were modeled.[35] This newly actualized and idealized fusion of body and machine produced swimming achievements that may never be equaled, and those who created this swimming cyborg soon had to come to terms with what this technoscientific hubris meant for the future of the sport.

For the 2004 Olympic Games, all the major manufacturers reengineered their suits. Adidas created a new Jetconcept, Arena produced the

Powerskin Xtreme, TYR manufactured the Aqua Shift, and Speedo designed the Fastskin FSII, each of which aimed, and in some cases claimed, to be an improvement over previous generations of suits. Arena declared its suit "the fastest and the most technologically advanced swimsuit ever made until then [by providing] superior muscle compression while still allowing optimal blood circulation, slower buildup of lactic acids, reduced muscle vibration, and unbeatable freedom of knee movement."[36] Speedo and TYR championed different approaches to swimsuit design. Speedo's computational fluid dynamics modeling refined the Fastskin FSII sharkskin design that the company used in prior sharkskin suits. Speedo determined that "friction drag constitutes up to 29 percent of a swimmer's drag . . . much more than the 10 percent previously thought."[37] They developed gender-differentiated suits that they claimed reduced passive drag by 3 percent and 4 percent for women and men, respectively. To address prior concerns about fit and comfort, Speedo developed "Flexskin," a stretchy fabric that served as the connective tissue to the "harder" surfaces found on the suit's torso. The Speedo Fastskin FSII suit also innovated "titanium-silicon scales on the inner forearm that grip the water better on down strokes [and] rubber bumps across the chest [to] help reduce another type of resistance called pressure drag."[38] TYR designed its Aqua Shift suit to increase frictional drag in order to decrease pressure drag (dictated by the shape of a swimmer's physique) and wave drag (created by waves in a pool). To achieve the designed drag effect, TYR built its suits with "three raised rings of equal height, called trip wires, placed where the circumference of the body is greatest—one around the calves, another around the buttocks, and the third around the chest."[39] TYR stated that its designed decreased drag by 10 percent.

In writing about the Speedo and TYR suits for *Scientific American* prior to the Olympic Games, Frank Vizard presciently wrote: "If all three spots on the winners' stand go to Speedo wearers, then . . . TYR's Aqua Shift suit . . . will become symbolic of its failed effort."[40] But Vizard's comments also indicate how comfortable the scientific press was with the impact the suits might have on the competition. At this point, they did not raise significant concerns. Speedo did upstage TYR's suit, but the suit's design was overshadowed significantly by the heroic and celebratory media coverage of Michael Phelps's eight-medal-winning performance in Athens. Though the development of the Speedo Fastskin FSII was more of an incremental technoscientific step than a monumental leap in suit design and performance, Phelps and this suit cemented the new swimming cyborg into the consciousness of the public. Thus it became hard to disentangle his achievements from the suit that he wore during his successes. In a sense, the media inescapably fused together Phelps's body and this new suit to fuel the public, technoscientific, and governmental fascination with fast suits and the performances that this union of body and machine could produce. At times, the heroic narrative emerging around Phelps seemed to transcend the suit and quell concerns that its technoscientific power had become vastly more important than the human body. As swimsuit technoscience matured, researchers produced new studies to conclusively show that these new fast suits unequivocally improve swimming performance.[41] Eventually, the suits and a technoscientific narrative consumed the heroic narrative and the athletes themselves, reducing swimming to a technoscientific competition rather than an athletic one.

Collaborations between swimsuit manufacturers and technoscientific institutions such as universities and industrial research laboratories were necessary to make the next revolutionary step in swimsuit design. Most companies began thinking about their suits for the 2008 Beijing Olympic Games in earnest after the 2004 Athens Olympic Games. Speedo assembled a team that would soon produce the LZR Racer, the suit that altered the swimming record books like no other. With the LZR Racer, it became apparent to many that the suits were not just benign technical equipment but well-thought-out technoscientific instruments specifically designed to significantly decrease drag and allow swimmers to swim faster than ever before. Speedo aptly named their new suit the Beijing suit (for the upcoming 2008 Olympic Games). Speedo's England-based research and development facility, Aqualab, led the development of this new suit. In January 2006, Aqualab's Jason Rance organized a meeting at the Australian Institute of Sport (AIS) with the major technical experts on the project.[42] Along with Rance, the group included Bruce Mason, who led the AIS's Aquatic Testing, Training and Research Unit; David Pease, a biomechanics expert at the University of Otago School of Physical Education, Sport and Exercise Sciences in New Zealand; Barry Bixler, an engineer at Honeywell Aerospace and an expert in computational fluid dynamics analysis who, as a Speedo consultant since the early 2000s, led the hydrodynamic design, analysis, and testing for the Fastskin FSII before the 2004

Athens Olympic Games; Rick Sharp, a physiology professor at Iowa State University who co-authored a paper concluding that the Fastskin FSII suit did not provide physical or psychological advantages to swimmers; and Steve Wilkinson, a NASA aerospace engineer at the Langley Research Center.[43]

This international team effectively lent their expertise to the design process and built on Speedo's history of collaborative work. Sharp made sure that the new suits did not appreciably diminish a swimmer's power and were reasonably comfortable. The suits needed to be hydrodynamically efficient, but athletes also needed to believe that they could perform well in them. In elite sports, confidence is paramount, and the suits needed to inspire trust. Bixler, with his computational fluid dynamics expertise (honed through his work with jet engines), focused on improving the process of gathering and interpreting data on the complex movements in swimming. By more precisely understanding key sets of variables that impact passive and active drag, the team could provide Speedo with the data needed to build a higher-performing suit. At NASA's Langley Research Center, Wilkinson tested multiple fabrics evaluated by Speedo. He studied almost sixty fabrics in their seven-by-nine-inch low-speed wind tunnel to determine "which fabrics and weaves had the lowest drag."[44] They eventually chose a fabric produced by the Italian textile manufacturer Mectex. Pease assessed multiple versions of the Beijing suit in a water flume at the University of Otago and oversaw four hundred hours of testing. At the AIS, Mason managed the direct testing of the suits by swimmers. This testing demanded a great deal of work from the swimmers. Trainers of Eamon Sullivan, one of Australia's strongest sprint swimmers, canceled the swimmer's second day of testing because of the extra wear and tear it put on his body.[45]

Not only was there a great deal of pride at stake but also a great deal of money. Swimsuit manufacturers had long subscribed to the American automobile racing mantra of "win on Sunday, sell on Monday." By 2008, the NPD Group, a market research firm, estimated the global market for elite and training swimwear at $515 million and the total swimwear market at $4.3 billion.[46] Thus winning in the pool with heroic and celebrated athletes did positively affect the bottom line. After admitting that the FSII had not been as revolutionary as initially hoped, Speedo aimed to design a suit that provided its sponsored athletes with the necessary competitive advantage. In the run-up to the 2008 Beijing Olympic Games, swimmers were going to have more than one option for a fast suit, and Speedo wanted to make sure that their suit set a new design and performance standard.

A brainstorming session took place in January 2006. In the early development stage, Aqualab considered all kinds of ideas, such as replicating "colors on the body and arms of the suit to improve hand-eye coordination."[47] Though this particular idea did not gain traction, it illustrates that Speedo was willing to pursue any good idea, no matter how outlandish. Nevertheless, drag reduction became the focus because, regardless of body shape, the reduction of friction between the swimmer and the water was the key technical problem to solve. In the summer of 2006, Mason received the first prototypes of the suit that would be known as the LZR Pulse.[48] Throughout the testing cycle, the time required to put on the suits became a significant concern. The technoscientific design mentality was to have bodies yield to the design and the shape of the suits. Swimmers fought for up to thirty minutes to twist, pull, and wiggle their bodies into the early suits. Redesigns eventually reduced the dressing time to fifteen or twenty minutes, but a side effect of fifteen minutes of pulling on the fabric was that the suits only had a handful of wears before degrading to a level where they were no longer effectively "fast." FINA approved the suit in January 2008, and Speedo launched the final version of the suit, the "LZR Racer," in February 2008 (figure 1.2).

The LZR Racer was somewhat of a tour de force in swimsuit design. Built primarily out of polyurethane and proprietary elastane nylon manufactured by the Portuguese-based company Petratex, the suit was welded together at every seam and juncture. It was a material representation of the best technoscience of the time. The suit compressed the body into a more idealized hydrodynamic form while having enough stretch to enable propulsive arm and leg movements. The hydrophobic fabric repelled water while simultaneously providing more buoyancy, allowing swimmers to ride higher in the water. Speedo claimed—as it had with all previous Olympic swimsuit introductions—that the new suit was faster than any other suit before it. Though multiple swimsuit manufacturers had been publicly attesting that their new Olympic suit was the best suit ever produced, Speedo's launch redefined the collective attitude and perception about corporate Olympic year swimsuit posturing. Speedo instituted a full media blitz on February 13, 2008. Prior to the official unveiling at Espace

Figure 1.2. Michael Phelps (*center*), Amanda Beard (*left*), and Natalie Coughlin in Speedo's LZR Racer suit, 2008. The combination of Phelps, the suit, and his 2008 Olympic performances transformed the technoscientific landscape of competitive swimming. Photo by Peter Foley; used by permission of epa european pressphoto agency b.v. / Alamy Stock Photo

(a high-profile media event venue in New York City), United States Olympic medal hopefuls Ryan Lochte, Katie Hoff, Amanda Beard, Michael Phelps, Natalie Coughlin, Kate Ziegler, and Dara Torres all appeared on NBC's *The Today Show* wearing the suit and touting how amazing it made them feel and perform. At the press conference a few hours later, the septet again displayed the suit as they all struck superhero poses.[49] Speedo's press release captured this otherworldly sentiment by quoting Phelps's excitement about the suit: "When I hit the water, I feel like a rocket. . . . I can't wait to race in it—this is going to take the sport of swimming to a new level."[50]

It is doubtful that Phelps recognized that his comments supported the tradition of comparing and reducing bodies to mechanical devices. His rocket descriptor supported the belief that technoscience was becoming the most important component of elite swimming. But comments like this also emphatically solidified Speedo's claim as the leading technoscientific developer of competition swimsuits. Yet it is unclear if Speedo realized that it was potentially opening a Pandora's box on the proliferation of technoscience in swimming by aggressively flaunting the performance of its suits to the world. Initially, this maneuver appeared to be more than a carefully orchestrated marketing strategy because athletes wearing LZR Racers dominated their various countries' Olympic trials. Speedo actually delivered on its claim of record-breaking performance, but the Beijing games proved its technoscientific superiority. Time would show that the suit was arguably too good. In a swimming world comfortable with incremental technoscientific advancements, the LZR Racer produced a monumental leap in swimming performance. The suit also became a flashpoint that inspired other companies to unleash their full energy into learning, acquiring, and developing the full complement of technoscientific power and skill to develop even faster suits. If FINA willingly approved the LZR Racer, there was no limit to what was now possible.

As athletes displayed the suit's overwhelming power, concerns about fairness and what the suit meant for the history and tradition of the sport again arose. One of the most prominent voices raising serious concerns about the suit was the Italian national swim team coach, Alberto Castagnetti. He did not hold back concern when he stated the following in late spring 2008: "This [technology] is going down a very dangerous road. . . . It removes the purely competitive aspect of the sport and puts outside factors into play. Swimming has always been based on ability. Now, there are other aspects. It's like technological doping. It's not in the spirit of the sport."[51] The claim of technological doping was not a benign way to categorize the suits. Castagnetti clearly intended to link the Speedo suits to the troubling narratives of performance-enhancing drugs, which were exploding in the world of cycling in 2008. He wanted to make the point that these suits were as powerful and as problematic as performance-enhancing substances.

For Castagnetti, these suits were no longer costumes or equipment but technoscientific devices that ripped out the moral heart of swimming. Castagnetti was not only arguing for the purity of his sport; he also was politicking for medals. The Italian national swim team was sponsored by Arena and wore their suit, the Powerskin R-Evolution. Though Arena noted that this suit was the lightest competition suit ever created (made from fabric instead of polyurethane), it did not measure up to the Speedo design. Moreover, with the Olympics on the horizon, it was not realistic for Arena to improve on its design for the games. Speedo may have been in the sights of Castagnetti and those concerned with what these suits would do for the public perception of the sport, but other manufacturers were more than ready to provide their swimmers with the best technoscientifically designed suits money and engineering talent could produce.

Even though Speedo claimed that the LZR Racer's ultrasonically welded seams reduced drag by 6 percent compared to a sewn seam, the new fabric reduced skin friction drag by 24 percent compared to the company's previous generation suit, and the suit's compression capabilities increased efficiency by 5 percent, it was in the pool that the suit cemented its technoscientific power.[52] By every quantifiable standard, the LZR Racer was unquestionably fast. As soon as athletes began wearing the suit in competition, world records began to fall. By April 4, 2008, athletes wearing the suit had claimed nineteen world records.[53] It became evident to everyone in swimming that the suit was an undeniable force. Being an Olympic year, there was a great deal at stake professionally for athletes, as well as those designing and engineering swimwear. Arena was the company most upset with the state of swimming. It voiced its displeasure with FINA because, as the sport's governing body, FINA authored the rules regarding permissible forms of technology within the sport and charged itself with the responsibility to test all equipment. On both accounts, Arena felt that FINA had underestimated the power of the LZR Racer and as a result was a poor steward for the sport.

On April 4, 2008, Arena's CEO, Cristiano Portas, submitted an open letter to FINA, delineating his concerns with the LZR Racer and with FINA. Portas noted the unprecedented leaps in performance by those wearing the suit and the "alleged buoyancy advantage" but coyly deflected his company's own self-interest in derailing the suit's use by indicating that "Arena, as a leading worldwide manufacturer of technologically advanced swimming equipment, has been asked by athletes, coaches, journalists and sports commentators to take a stand on this issue." As his statement proceeded, he honed in on what he saw as a crisis of athletic credibility, and FINA's role in the creation of that crisis. He called for FINA to "scrutinize and reaffirm the effectiveness of its policies," because if the governing institution did not embrace its role as a regulator the sport would become a farce as an athletic competition. Specifically, he demanded that FINA enforce its own rules requiring suits to be made from "regular flat fabrics [with] no outside applications . . . added." But most important, he wanted FINA to uphold rule SW 10.7, which stated that "no swimmer shall be permitted to use or wear any device that may aid his speed, buoyancy or endurance during a competition." The letter ended with a fairly direct threat: "Should FINA decline effective intervention. . . . Arena would have no alternative but to pursue in every appropriate forum all available remedies to achieve a final and satisfactory clarification."[54]

Though Portas was angling for his company's own interests, he was right in most regards. Even to those outside the sport, the Speedo LZR Racer was a game changer. In most years, the technoscientific power of Speedo's swimsuits may have gone unnoticed, but this suit was launched in an Olympic year, when nationalism draws more attention to sports. Portas's letter also came on the eve of FINA's short course world championships, which were scheduled for April 9–13 in Manchester, England. An additional wrinkle in the story is that Stephen Rubin, the chairman of the Pentland Group, of which Speedo is a subsidiary, was the head of the organizing committee for the Manchester world championships. For many, this arrangement reeked of conflict of interest. It did not help that after the championships, Rubin attested that Speedo "always played by the rules [and] as far as we're concerned, there is nothing wrong with our swimsuits, and it . . . conforms with FINA's rules."[55] This only enhanced the perception that FINA gave Speedo preferential treatment.

Competing manufacturers such as Adidas and Diana (whose new design was denied by FINA in 2007) believed that the LZR Racer's "multiple layers and use of neoprene aid buoyancy and [that] the suit's rubber panels violate FINA's laws on materials." Arena's Giuseppe Musciacchio conveyed a growing sense of hypocrisy when he indicated: "We also had a lot of ideas on the table but we could not use them because FINA would not accept them. . . . We do not have to accept this situation just because FINA made a mistake. It seems that if you don't respect the rules, you are an innovator—where are the ethics here?" This last comment is telling because many felt that Speedo and FINA had agreed to change the terms of the game without notifying anyone else until it was too late. Musciacchio did not mince words when he demanded that it was time to "stop now and test these suits. . . . We are always pushing technology but if we do not have rules, we will have anarchy."[56]

This set of circumstances drove Speedo's competitors to focus on the process of swimsuit approval. Before the 2008 short course world championships, FINA attempted to manage its central role in the LZR Racer controversy as other suit manufacturers began to publicly question the efficacy of suit evaluation. As a preemptive response to the ensuing questions, FINA issued the following statement specifically directed at Arena's open letter of April 4, 2008:

> All swimsuit manufacturers are very important partners to FINA, Member Federations, swimmers, coaches and officials, substantially contributing to the growth of our Sport.
>
> Consequently, FINA, in partnership with the manufacturers, implemented the following:
>
> 1. Regulations on swimwear approval have been formulated in close co-operation with the manufacturers, whose participation and contribution was essential;
> 2. The FINA swimwear approval process has been implemented since 21 November 2005 without any subsequent specific remarks made by the manufacturers on the process;
> 3. FINA is always willing to examine issues in connection with the swimsuit approval. However, to the best of our knowledge, there is no objective scientific evidence on the alleged buoyancy advantage provided by 'SPEEDO LZR Racer', 'TYR Tracer Light' or any other swimsuit approved by FINA;
> 4. On 12 February 2008 (i.e. before any discussion on swimsuit arose), FINA invited all the swimwear manufacturers to participate in a meeting on 12 April 2008 in Manchester (GBR) on the occasion of the 9th FINA World Swimming Championships (25m). The stated purpose of this meeting is to review and update, if considered necessary, the procedure and requirements for swimwear approval;
> 5. We underline that at FINA competitions, the rule GR 5.6—"the manufacturers must ensure that the approved new swimsuit will be available for all competitors"—will apply.[57]

The numbered points staked out FINA's position that it had done nothing wrong. It noted that it used the same collaborative process for evaluating suits since November 2005 without any prior complaints. FINA also indicated that it had nothing to hide. It reminded readers that it scheduled the meeting with swimwear manufacturers in early February and dismissed the allegations about the LZR Racer's buoyancy advantage based on a lack of "objective scientific evidence."

Clearly, all was not well within FINA. One reading of the situation was that FINA called a meeting with the manufacturers to discuss suit designs because it knew what was coming. Speedo released the LZR Racer on February 13, 2008, one day after FINA called for the meeting. If FINA did evaluate the suit in "close co-operation" with Speedo, it probably had a fairly decent understanding of the performance capability of the suit. Arguably, FINA felt that it needed to hold a meeting because the technology had moved beyond its capability to regulate it. Many of the complaints about the LZR Racer stemmed from the belief that it trapped air, increased buoyancy, and thus allowed athletes to swim faster by riding higher in the water. FINA had no way to dispell these concerns because it did not have a standardized test for buoyancy, nor had it deployed the scientific analytics to standardize the way it tested and evaluated suit design and performance besides stating that it had used the same process since 2005.

At the conclusion of the April meeting, FINA, in a carefully worded press release, noted that the meeting provided an opportunity for it to receive suggestions from manufacturers regarding the approval process. Many manufacturers hoped that the LZR Racer would get shelved until more data was available on the suit, but FINA confirmed its legality by blandly stating: "All the swimsuits approved so far are complying with the specification." The most important component of the meeting centered on the fabric. FINA ushered in the technoscientific future by stating the following: "In regards to the swimwear material, the discussion clarified that there was a broad understanding between the manufacturers and FINA that the rules were not meant and should not be interpreted as limiting the materials to fabrics stricto sensu but that other material could be used."[58] Manufacturers interpreted this statement as a mild acceptance that technoscience would drive future competitive swimsuit design. Yet for most manufacturers, it was too late to redesign their suits for the Beijing Olympics.

The fact that FINA allowed swimmers to use the LZR Racer for the Olympics created massive tensions around the demand and pressure to wear the suit. In April 2008, the head coach of USA swimming, Mark

Schubert, cogently spoke about the moral and ethical challenges athletes faced regarding suit choice.

> My advice to athletes is "you have a black and white decision—the money or the gold medal." And it's going to be a real test of character as to what choice they make. There is no doubt the suit makes a difference and there is no doubt that there is one manufacturer that's put millions into research while the other manufacturers are more into fashion. If you take best times of world record holders and their new times, the difference is 2 per cent. Nobody at this level . . . can afford to give up 2 per cent. It is not rocket science.[59]

As all of the disputes around what to do escalated, Schubert's comments cut through this tension. Non-Speedo athletes were going to have to choose between lucrative sponsorship deals or the desire to win a medal. More importantly, Schubert did not see anything wrong with what was happening. Since the United States swam in Speedo suits, he was more than happy with the way things seemed to be turning out. He was also proud of the work that Speedo and the athletes put into developing the suits.

> I think really the latest three generations of suits have had a great effect. I'm very excited about the technology. I think it's another exciting aspect of this sport. . . . I went to Canberra to watch Michael Phelps, Ryan Lochte, Natalie Coughlin, Kate Ziegler and Katie Hoff and watched them do eight hours of testing with the new suit. . . . Speedo [has] been working with athletes and getting feedback from them, asking them how they feel. It's very important to get that input from the people who have to wear the suit, telling the manufacturers how they felt about the suit and how it helps and how it doesn't help.[60]

It was clear to him that Speedo's competitors could not compete. By June, when most countries hosted their Olympic trials, the LZR Racer transformation was in full swing. Though Canada and Italy prohibited its use in their Olympic trials, most federations and competing companies acquiesced. The Federazione Italiana Nuoto, Italy's swimming federation, who had a contractual agreement to wear Arena suits, eventually agreed to allow its swimmers to wear the LZR Racer as long as they were willing to pay a fine. Nike initially allowed its sponsored athletes to use the suit in the Olympic trials, and it extended that experiment to the Beijing Olympics. Japanese swimmer Kosuke Kitajima's pre-Olympics racing exemplified the tension around suit choice. At the Japan Open swim meet on June 8, 2008, Kitajima shattered the existing 200 m breaststroke world record by nearly a second in one of his first competitive events wearing the LZR Racer. Nihon Suiei Renmei, Japan's swimming federation, similar to other national swimming federations, begrudgingly agreed to allow its athletes to wear Speedo suits and break agreements with its corporate sponsors, Mizuno, Asics, and Descente. Schubert was correct in that competitors did not catch up and the swimmers voted with their desire to win.

By the opening of the Olympics on August 8, 2008, the LZR Racer had established itself as the technoscientific tool of choice for all swimming competitions. From a technoscientific perspective, it was the most iconic symbol, both visually and materially, of the radical shift happening within the world of sport at that moment. As the swimming events commenced, most media outlets interlaced the reporting of the events with commentaries about the choice of suit. In many instances, the reporting was as much about the suit—the time it took to get it on, the materials in use, and the controversies over its use—as the athletes themselves. The suits provided a productive narrative arc for the swimming competitions. This attention did raise public concerns over what really was doing the winning. The heightened media discussions about the power and efficacy of the suit tacitly made the strong argument that the suit overrode physical ability. Near the end of the Olympics, there was a bit of backtracking through the celebrated eight-gold-medal performance of Michael Phelps. The revised narratives argued that everyone had access to the same suit, thus the playing field was level. With this reasoning, Phelps was the "best" swimmer because the conditions were the same for all competitors. The same cannot be said for the record books, however.

Athletes wearing the LZR Racer suit won an astounding 94 percent of all races at the 2008 Olympic Games. The winner of each men's race wore the suit. Athletes wearing the LZR Racer took home 89 percent of all medals available and only two Olympic records were left standing by the end of the Games.[61] The suit unquestionably had changed swimming and its record books, but it also changed the future of suit design. By allowing athletes to wear the extremely fast LZR Racer, FINA licensed companies to unfurl their research teams to bring an increased level of technoscience to the design of their products. Some companies, such as Nike, left the sport for the short term. Others, such as Jaked and Arena, took up the challenge. Jaked and Arena were two companies that moved beyond the LZR Racer and its application of polyurethane panels to the sections of a swimmer's body that create the most drag. The Jaked J01 and the Arena X-Glide suits were made completely out of polyurethane. These companies created amazingly efficient seamless suits by thermally welding polyurethane panels together. Speedo showed how potent polyurethane could be, so other companies took the material to its logical conclusions and built suits completely out of it. The next year and a half was probably the most challenging period in FINA's history. It had opened a technoscientific floodgate and was struggling to throttle the deluge of suits pushing the limits of the sport. This swift rate of change highlighted FINA's inability to regulate and determine which suits were and were not permissible.

In light of this turmoil, FINA announced on December 22, 2008, that it would again host a meeting with swimwear manufacturers to discuss swimsuit "material, thickness, use, shape and availability." This was a broad meeting at which "representatives of the FINA Technical Swimming Committee as well as FINA Athletes, Coaches and Legal Commission" were to be present.[62] This meeting, scheduled for February 20, 2009, along with comments and reports presented at "an elite coaches' forum" in Singapore in January would inform suit policy decisions that FINA planned to develop for its bureau (its elected governing board) meeting from March 12 to March 14, 2009, in Dubai.[63] The goal was to get out in front of the next generation of technoscientific suits that most certainly were on the horizon. Ideally, they would gather relevant data from manufacturers, trainers, athletes, and, most importantly, scientists to guide their decision-making process.

FINA's choice to schedule this meeting was prompted by collective protests at the 2008 European short course championships held in Croatia from December 11 to December 14. Fifteen of the top seventeen ranked European swimming federations signed a letter of protest regarding how poorly FINA handled the swimsuit issue. By the conclusion of the championships, 105 world records had been eclipsed since the introduction of the LZR Racer. The attitude that "average" swimmers could potentially challenge for medals and championships so long as they wore the "right" suit was rising. The massive turnover within the record books gave credence to this sentiment. Britain's swim coach, Dennis Pursley, summed up the predominant concerns when he stated that "it demeans the records and kind of cheapens them to an extent. . . . Up to this generation of suits I think it's just been maximizing performance but I think we're crossing the line to enhancing performance and to me that is a whole different area."[64] FINA's lack of decisive action delineating the differences between an acceptable suit/costume and an illegal technoscientifically designed, performance-enhancing device drew much of the ire.

At the February 20, 2009, meeting, representatives from sixteen swimsuit manufacturers and multiple FINA commissions (executive, technical, legal, coaches, and athletes) developed an agenda for the March 2009 FINA bureau meeting. The collected participants decided that design, material, buoyancy, construction, customization, use, control, and approval were the key areas of focus for a revised set of regulations. These basic guidelines were specific and important. In terms of design, the group determined that suits "shall not cover the neck and shall not extend past the shoulders nor past the ankles." Clearly, research showed that bodily flesh was an issue, so incasing the body more meant increasing bodily control through the suit. As for materials, they came to three conclusions: "the material used shall have a maximum thickness of 1mm; when used, the material shall follow the body shape; [and] the application of different materials shall not create air trapping effects." The first decision about thickness addressed the cellular structure of polyurethane and its ability to trap air. The thicker the polyurethane, the more air it could trap and the larger advantage it could provide a swimmer. About body shape, fast suits were designed to reconfigure bodies into more efficient shapes. Thus the rigidness of fast suits was to be eliminated. Finally, trapping air relates to concerns regarding buoyancy, which was a major complaint about the LZR Racer. FINA determined that suits "shall not have a buoyancy effect of more than 1 Newton" and that a "swimmer can only wear one swimsuit at a time" in order to avoid adding a magnitude of buoyancy for each additional suit worn. FINA also prohibited "any system providing external stimulation or influence of any form." One-off suits specifically modified for a specific swimmer that varied from the approved sample were also to be eliminated. Finally, FINA instituted a more rigorous and scientifically based control and approval system. It employed the services of Professor Jan-Anders Månson from the Swiss Federal Institute of Technology (EPFL) and Laboratory of Polymer and Composite Technology to "establish its own independent control/testing programme." FINA gave manufacturers until March 31, 2009, to submit their suits for examination and approval. FINA president Mustapha Larfaoui surmised: "With these amendments, FINA shows that it continues to monitor the evolution of the sport's equipment with the main objective of keeping the integrity of sport. While we need to remain open to evolution, the most important factors must be the athletes' preparation and physical condition on achieving their performance."[65]

FINA intended the Dubai meeting to be a turning point, and for a few months it was. From this meeting, the Dubai Charter emerged, which in many ways institutionalized the structures delineated at the February 20, 2009, meeting. The charter aimed squarely at rapid technoscientific evolution within swimming and reminded those both within and outside this sporting culture that "the main and core principle is that swimming is a sport essentially based on the physical performance of the athlete. This is the fundament which FINA has and will continue to preserve as its main objective and priority. FINA brings together athletes from around the world to compete on equal conditions and thereby decides the winner by the athlete who is physically the best." Yet as FINA critiqued technoscience's reach into its sport, it did not want to be seen as technophobic. The charter expressed its belief "that swimming, like all other sports, should . . . integrate the natural progress and improvements in technology where this helps, bettering the conditions under which the athletes compete and optimising their athletic performances." This was an extremely difficult tightrope to walk. On one hand, it sought to delimit technology in sport while, on the other hand, embrace its potential. FINA's solution in 2009 was to bolster its requirement for approving swimwear.[66]

In addition to the requirements drafted in February 2009, FINA also developed a four-step approval process. First, the approval process required that "any swimwear used in FINA competitions and Olympic Games shall comply with these new rules and shall be a model approved by FINA in accordance with these new procedures." Second, it instituted a specific timeline for suit approval to end the historic procedure of a company launching its new suit within months or weeks of a major competition. The first date was November 1, 2009, and August 1 thereafter, and if approved the suit must be "available on the market at least 6 months prior to the forthcoming FINA World Championships or Olympic Games." Third, FINA published a publicly available list of approved swimsuit models. And finally, FINA used a "permeability value test" to evaluate the porousness of swimsuits, which proved to be the most transformative of the added requirements. It specifically noted that "non-permeable material can only be used for a maximum 50% of the total surface of the swimsuit for full-body models. For these models, the maximum surface of non-permeable material to be used on the upper and lower part of the swimsuit shall be respectively 25% on each part. Non-permeable material shall be distinguishable." FINA directed this regulation at the accusations and concerns around air trapped within suits creating buoyancy. FINA's efforts initially assuaged many concerns, but only for a short time.[67]

FINA appeared to have resolved many of the issues and tensions by May 2009, when it released its "2009 List of Approved Swimsuits." The list included suits from Adidas, Agon, Akron, Arena, Asics, Descente, Diana, Finis, Jaked, Kiwami, Mizuno, Orca, Rocket Science Sports, Sailfish, Speedo, Sports Hig, TYR, and Yamamotohokosyo Corporation. Each major manufacturer who wanted a suit approved needed to receive a FINA stamp of approval. Of the 348 suits tested by the group led by Professor Jan-Anders Månson at the Swiss Federal Institute of Technology, only 202 received approval. FINA banned 10 outright for "not passing the tests of buoyancy and/or thickness," while the remaining 136 were provisionally sidelined until the manufacturers altered the suits to fully conform to the Dubai Charter ruling requiring that the suit "material shall not be constructed to or include elements/systems which create air/water trapping effects during use." FINA allowed the companies thirty days to alter their suits and resubmit them for reevaluation.[68] Unfortunately, thirty days was not long enough to resolve the issue, and by the time FINA's executive committee met on June 19, 2009, the recommendation transitioned from a scientifically based regulatory authorization to an economically based insider concession.

FINA backtracked on its demands regarding the elimination of air being trapped in suits because it "found that the evidence of 'in use' air trapping effect is complex and that it would require considerable time to create and implement comprehensive control mechanisms and test methods which would permit to establish the effect with absolute certainty in connection with particular swimsuits."[69] It was quite optimistic that a scientific protocol regarding trapped air, satisfying all invested in the process, could have been developed in a month. FINA appeared to be equally interested in the negative financial impact of what banning suits meant for the various manufacturers who supported the institution and infrastructure of swimming. In fact, it indicated that in situations "where there was a reasonable doubt as to the effect" of air and, subsequently, buoyancy, FINA deemed it "important to take into account that the outcome of the approval process can have substantial impact on existing production and stocks."[70] FINA fully understood its symbiotic relationships with swimwear manufacturers and was mindful of the fiscal strain it could create throughout the various swimming networks if it banned a certain set of suits. In this June 22, 2009, press release, FINA wrote about it being a "transitory" period and that it would continue working toward better evaluative methods and planned to develop and disseminate new criteria for 2010. The result was that FINA, at least temporarily, banned nearly nothing, and by adding 186 new suits it ballooned the approved suit list to 388. Clearly, it did not have control of the situation or a strong plan to manage the emerging technoscience.

Jaked, on the same day FINA released its list of suits, announced that "FINA has attested that the swimsuit J01 . . . fully complies with the regulations in force [and that] Jaked proved on . . . scientific indisputable evidence that the swimsuit J01 does not provide or take any advantage from air-trapping and that its special composition makes it totally permeable to air at any condition, even when not stretched." Yet as pleased as it was with the verdict, the company did take a jab at FINA by stating: "We regret the prejudice that this affair unfairly brought to the reputation and brand image of Jaked and the damages caused by the loss on sales of our product J01 prior to the World Championships, compromising a potential huge and quick spread of its diffusion."[71] This last comment made it abundantly clear that Jaked felt the evolving rules had a negative impact on its bottom line. But all was not fine for everyone.

TYR's suit, the Tracer B8, was one of the few that remained banned. French swimmers Amaury Leveaux and Aurore Mongel appealed to the CAS on the grounds that FINA had released similar and previously banned suits such as Arena's X-Glide and Jaked's J01, and that disallowing them to wear the TYR suit would inflict irreparable harm on them at the upcoming world championships.[72] So even after it opened the floodgates for suits, FINA was still dealing with a handful of issues. By mid-July, it was under mounting pressure to reclaim the sport from this runaway technoscience. USA swimming's executive director Chuck Wielgus noted that record-breaking performances had become so commonplace that instead of excitement and awe, new records received "a golf clap—just a polite applause." He continued to note that new world records are "not as special and it just raises the expectations for the athletes."[73]

All of these tensions played out during the 2009 World Aquatic Championships. Two days before the swimming events commenced, the FINA congress once again changed the rules regarding swimsuits. But this time they chose to make a strong statement against the technoscience of swimsuit design. The key rule change centered on materials, in that it agreed to return to "textile fabric." FINA defined textile fabric as "material consisting of natural and/or synthetic, individual and non consolidated yarns used to constitute a fabric by weaving, knitting, and/or braiding."[74] This ruling was a fundamental shift away from the technoscientifically designed hydrodynamic suits and polyurethane. FINA made a strong statement about how it wanted to return the sport to its athletes. This should not come as a surprise, because, as Chuck Wielgus noted, the breaking of a new swimming world record was becoming so commonplace that the occurrence had become mundane.

FINA certainly understood that the suits sublimated celebrated and marketable athletes. It managed athletic competitions, not technoscientific design contests. It pushed even further by legislating that "for men, the swimsuit shall not extend above the navel nor below the knee, and for women, shall not cover the neck, extend past the shoulder, nor extend below the knee. Furthermore, no zippers or other fastening system is allowed." This ruling was not only aimed at the hydrodynamics of the full-body swimsuits, but it was also directed at destabilizing the robotic look of the full-body suits. Substantially decreasing the amount of fabric that covered and disciplined swimmers' bodies potentially slowed the swimmer down. But the ruling also changed the visual representation of swimmers to the public. The showing of more skin and flesh visually represented a return to the primacy of the body and FINA's desire to rein in the runaway technoscience. FINA also noted that it would invert the power of technoscientific testing to more effectively regulate swimsuit design, manufacturing, and materials. In this regard, FINA indicated that "only measurable scientific tests will be performed within the frame of the swimwear approval procedure." Smartly, they moved away from what some thought of as a subjective approval process, but this also allowed them to sidestep their responsibility in the promotion of the swimsuit arms race.[75]

FINA also ruled that "for thickness, the 1mm limit will be adjusted to 0.8mm, for buoyancy the present value of 1 Newton will be reduced to 0.5 (FINA will even consider the limit of 0 Newton), and for permeability the material(s) used must have at any point a value of more than $80l/m^2/$ second. Permeability values are measured on material with a standard multidirectional stretch of 25%."[76] By decreasing the technoscientific specifications for swimsuits, FINA definitively moved to give swimming back to the swimmers. In many ways, this move was earnest in intent but an illusion in practice. FINA had already let the swimsuit designers free to play, create, and invent, which is generally impossible to stop. The governing body met again in Rome on July 31 to finalize its revamped rules and structure around swimsuit design. At this meeting, FINA also marked the end of this era of technoscientific suit design by confirming that the new rules would go into effect on January 1, 2010. This enabled suit manufacturers to redesign suits for the new standard, but it also allowed competitors to swim in the most recent version of fast suits approved on June 19, 2009, until the end of that calendar year.[77]

On January 1, 2010, FINA metaphorically and materially wrested the sport away from the suit manufacturers, material scientists, and a technoscientific momentum that fundamentally altered the public image of the sport. But as FINA attempted to realign the sporting cultural equilibrium, it was clear that the sport had morphed into a different configuration. Specifically, the suits had restructured the sport's record books. These key metrics, used to compare swimmers across generations, could not be used in this same manner. In the end, FINA allowed the records to stand. The records existed as a cautionary reminder of the massive shift technoscience precipitated within swimming. By the close of the 2016 Olympics, only six (two of which came at these games) of the 20 men's world records set before January 1, 2010, had been eclipsed, whereas, on the women's side, 13 (six at the 2016 Olympics) of the 20 world records had been surpassed. Though these numbers may call into question the power of the suits, these years also saw some of the historically lowest rates of world record breaking on both the men's and women's sides of the sport.[78]

Overall, the public, most of whom did not follow swimming outside of the Olympics or select national and world championships, did not conceptually understand the depth of what was at stake for the sport of swimming. As researchers began to publish studies on the effectiveness of the suits, it became abundantly clear that technoscience had become the driving force behind years of increased athletic performances. As it stands, it will probably take several more years to adequately assess and fully grasp how fast suit technoscience impacted this moment in swimming history. FINA, though slow to react initially, decisively acted to discipline the sport's technoscience. In theory, the FINA decision returned the sport to the swimmers, the in-pool competitions, and the belief that unaugmented athletic bodies should determine the outcomes of sporting competitions. Thus FINA, in its power to govern competitive swimming, chose to reprivilege athletic bodies over technoscience.

2

Gearing Up for the Game
Equipment as a Shaper of Sport

In contemporary sport, equipment is central. Whether it is uniforms, shoes, gloves, bats, balls, or other more complex devices, all athletes play with or use some form of technoscientific gear. Sport governing bodies have a vested interest in managing the use of technoscientific equipment that challenges a sport's cherished versions of authenticity. But publics, athletes, and equipment manufacturers are equally invested in similar forms of authenticity. These investments can make entire sporting cultures disregard the impacts of technoscience to the point that game-changing gear does not cause a sport to reevaluate its history and traditions. Technoscience, once seen as making sport safer and more exciting as well as extending the capabilities of athletes, can be construed as devices that undermine the integrity of sport. From the clubs that the United States Golf Association (USGA) deemed illegal because they enabled golf balls to fly too far, to the Fédération Internationale de l'Automobile's (FIA) ban of driver aids such as traction control in Formula 1 racing, sport governing bodies increasingly have attempted to legislate against existing, new, and emerging technoscience.[1]

This chapter illustrates the ways in which sporting cultures choose to ignore, dismiss, or legislate away the power of technoscientific equipment central to a sport from its origin. Though there are multiple sporting environments to explore these relationships, three illustrative examples—athletic shoes, synthetic basketballs, and racing bicycles—will tease out the intricate ways in which sporting communities struggle to narrate technoscience as instrumental. As we saw in chapter 1, it was clear to most everyone that new fast suits were decidedly different from suits of previous generations and that their use altered the competition like nothing before. However, the examples presented here explore situations in which the transformative impacts of the technoscience were not so clear-cut to the sport governing bodies, the athletes, the fans, or those making the equipment. The reconfiguration of existing sporting equipment destabilized how sporting cultures understood their relationship with their given sport's technoscience. These examples reveal how facets of sporting communities differentially understand, address, and package the ways in which externalized equipment impacts their sport. Specifically, these cases make the argument that all technoscientific equipment influences the outcomes of sporting competitions, no matter how hard sporting cultures strive to maintain narratives in which physical ability trumps technoscientific innovation.

It's Gotta be the Shoes

At the beginning of the 1954 World Cup soccer tournament, the Hungarian national team was the undisputed favorite. In the run-up to the tournament, the team won thirty-two straight games and only seemed to be getting stronger. The Hungarians made this abundantly clear in the previous year by soundly defeating England 6-3 at Wembley Stadium. Arguably one of the most significant soccer matches up to that point, the loss broke a fifty-two-year home-winning streak for the English. Three weeks before the start of the World Cup, the Hungarians faced the English again. On this occasion, the Hungarians engineered an even more convincing, 7-1, victory to cement their place as the unmatchable favorite. In group play, the Hungarians displayed their dominance and did not come close to losing a game. In the knockout stage, Hungary continued its brilliance, even though the team's star player, Ferenc Puskás, did not play due to a hairline ankle fracture he acquired in the round play against West Germany.[2]

The World Cup final was a rematch of a first-round game that West Germany lost 8-3. (It should be noted that West Germany fielded a reserve team for the first match.) Puskás returned for the final against West

Germany, and early in the game it appeared that it would be another rout. Eight minutes into the match, the West Germans were down two goals. By halftime, the West German team leveled the game at two goals each. Finally, in the eighty-fourth minute of play, Helmut Rahn scored the final goal, enabling the West Germans to overcome seemingly insurmountable odds and win the 1954 World Cup. In West Germany, the game became known as "Das Wunder von Bern." The Miracle of Bern is seen as a transformative event that ushered in a new collective West German identity.[3] The story of sport as a force of cultural transformation is familiar. The 1954 World Cup displays sport's role as a competitive space reinforcing national pride, tradition, and identity. Yet there is another narrative about this World Cup and West Germany that is equally as important: the story of technoscience. Not only was this the first televised World Cup soccer final, but it potentially was also the first decided by a calculated deployment of technoscience in the soccer shoes worn by the West German team.

This story begins in the 1920s, when Adolf "Adi" Dassler made his first pair of athletic shoes (figure 2.1). Through shrewd management, he and his brother, Rudolf, built their shoe company into one of the world's largest sports shoe manufacturers by the end of the 1930s. The Dasslers also understood the importance of aligning with celebrated athletes and having these athletes wear their shoes during significant sporting events. For instance, Adi convinced Jesse Owens to wear the brothers' shoes during the 1936 Olympics.[4] The company did not produce athletic shoes during World War II, but in 1947, after the war was over, they relaunched the company. A rift between the brothers resulted in a less-than-amicable split, which led Adi to create "Adolf Dassler adidas Sportschuhfabrik" (Adidas) in 1949, while Rudolf started Puma.[5] Adi's zeal for producing groundbreaking athletic footwear led him to design a new innovative soccer shoe for the West German team for the 1954 World Cup.

The story of Adi's transformative shoes usually begins with the weather. A forecast of rain for the highly anticipated July 4 final at Wankdorf Stadium was far from ideal. By the start of the second half, a steady rain had turned the field into a waterlogged quagmire. Traditional soccer shoes, with leather uppers and soles, were known to turn into slippery lead boots in the rain. Dassler had a plan for this occasion, made possible by his close friendship with the West German coach/manager, Sepp Herberger. Dassler designed and manufactured a pair of shoes for each West German

Figure 2.1. Adolf "Adi" Dassler in the midst of testing different cleats and configurations to optimize his interchangeable soccer cleat design, 1954. Photo by Hanns Hubmann, from Ullstein Bild collection; used by permission of Getty Images.

player out of thinner, lighter leather. But the true innovation was the sole. The shoes had removable cleats that could be exchanged, depending on the field conditions. At halftime, Dassler installed longer cleats, designed to effectively grip the quickly degrading soft and muddy turf. As the second half progressed, the West German team began to outplay the Hungarians, who struggled against their better-equipped opponents. Both World Cup and shoe history were made when Rahn's score put the West Germans ahead for good.[6]

The story of West German triumph through technoscience needed a willing audience and a savvy promoter. Clearly, Adi Dassler had a vested interest in designing the shoes that would propel the West German team to victory. This narrative gained traction quickly and launched Adidas as the shoe, and eventually the brand, in which serious soccer players competed. The simultaneous rise of West Germany as a soccer powerhouse only fueled this perception. But does this story ring true? Or, more specifically, did the shoes really influence the outcome of the game? This, like other stories of game-changing technoscience, is up for lengthy debate.

Film clips of the final match do not fully substantiate the descriptions of the West Germans running circles around the slipping and sliding Hungarians. Nevertheless, if one believes that the shoes did improve traction, the central issue is, how much did the effects of these devices influence the final outcome of the game? This is a key question because this technoscientific narrative of sport triumph would have never been created if the Hungarian team had won the World Cup in 1954. But it is significantly difficult to determine how much the shoes influenced game play. The game of soccer at the elite level is a beautiful game driven by individual and collective creativity and ingenuity with and without the ball. Though this skill is channeled through bodily movements and, ultimately, soccer-cleat-clad feet, it is incredibly difficult to discern the impact of the shoes. In fact, some Hungarian fans may feel that Welsh referee Mervyn Griffiths played a larger role in the final outcome of the game.

Griffiths was one of two linesmen in the final. In the closing minutes of the game, Puskás, and the Hungarian supporters, believed he scored the equalizing goal. Griffiths derailed their celebration when he raised his linesman's flag, indicating an offside call against Puskás, thereby nullifying the goal. The quality of this call has been debated ever since.[7] The way the game finished may partially explain why there was not a large outcry about the shoes. Hungarian supporters did not feel that the West Germans used an unfair secret weapon. They blamed the loss on Griffiths and his poor refereeing. The technoscientific twists and turns of this championship final received renewed focus in 2004, when documentary filmmakers Guido Knopp and Sebastian Dehnhardt released *The World Cup Wonder of Bern 1954: A True Story*. The film raised questions regarding whether the West German team received injections of performance-enhancing substances at halftime instead of the publicly declared vitamin C boosts.[8]

Both the technoscientific and cultural narratives of the 1954 World Cup are wonderful stories of West German redemption through sport. Equally important, however, was the distance they created from the long shadow of the Nazi era. The potential power of the shoes could not overcome the more important and valuable perspective that a renewed West

Germany rose to the challenge, brushed off its war wounds, and emerged as a world champion through its collective belief and hard work rather than through an innovative technoscientific maneuver. It is equally hard to imagine that Adi Dassler and his shoes could have supplanted the power of this desirable national message. The potential technoscientific potency of the shoes did not fit the larger culture's need to claim a reformed national identity emanating from the collective will of the people and not from a technoscientific twist.

When compared to today's sporting moments, these events seem almost quaint. The idea that a team would reveal a game-changing techno-scientific tool at the highest level of a sporting competition, at halftime, is almost comical. This occurrence probably would not be perceived as a brilliant tactical decision, but as a deceitful effort to undermine the purity of the game or competition, the power and authority of the governing body, and the trust of the public, resulting in accusations of cheating and multiple forms of protest. Nevertheless, it still is surprising that shoes, one of the most essential pieces of athletic equipment, rarely have been in the crosshairs of sport's regulatory infrastructure. Though in the 1980s, the National Basketball Association determined that Nike basketball shoes worn by Michael Jordan, while conforming to its performance standards, did not adhere to its aesthetic rules, namely color. This rule, which was eventually changed, became one of the most media-worthy instances of a sport governing body banning athletic equipment.

The popularized narrative of this event notes that the National Basketball Association banned the first Air Jordan version—the now-iconic "Jordan 1" (released on September 15, 1985)—and fined Jordan for wearing the shoes. NBA commissioner David Stern announced that the shoe's Chicago Bulls–inspired black-and-red colorway violated the NBA's uniformity rule, which required that "a player . . . wear shoes that not only matched their uniforms, but matched the shoes worn by their teammates."[9] It did not help the NBA that the black-and-red "Jordans" quickly became one of the most coveted athletic shoes ever, and the attention the NBA showered on the shoes only increased their desirability and marketability. Nike and Michael Jordan capitalized on the NBA's effort to rein in this form of aesthetic creativity. As the narrative goes, Jordan received a $5,000 fine each time he wore the shoes. The amazing publicity the media provided was well worth the fines, which Nike gladly paid. As the focus on the color of the shoes diminished, Nike smartly continued to promote the shoes and the Jordan brand to the level that in 2014, it "commanded 58% market share of the $4.2 billion U.S. basketball shoe market," which helped make Michael Jordan, one of only a handful of former athletes, a billionaire.[10]

It is unclear whether the NBA actually did ban any shoe or institute a $5,000 fine. Sneaker aficionados contend that the NBA did not question Jordan's wearing of the Air Jordan 1 but instead a black-and-red version of the much-lesser-known and non-Jordan-branded Nike Air Ship model that he wore during his rookie campaign in the 1984–1985 season.[11] Moreover, a February 25, 1985, letter sent from the NBA's executive vice president Russ Granik to Nike vice president Rob Strasser notes that "the National Basketball Association's rules and procedures prohibit the wearing of certain red and black NIKE basketball shoes by Chicago Bulls player Michael Jordan on or around October 18, 1984."[12] This brief letter to Nike does not imply a fine, sanction, or any other reprimand for wearing the shoes, although Howard "H" White, the vice president of Jordan Footwear, has remarked that the NBA "threatened to fine Jordan $5,000 for every game he played wearing them."[13] Nike reacted quickly and turned the NBA's admonishment into an advertising campaign. Nike latched onto the NBA's reprimand for wearing the black-and-red colorway to launch the first Air Jordan 1 in the fall of 1985. The associated television commercial for the Air Jordan 1 showed Jordan handling and dribbling a basketball while the narrator stated: "On October 15th, Nike created a revolutionary new basketball shoe. On October 18th, the NBA threw them out of the game. Fortunately, the NBA can't keep you from wearing them."[14] History has shown that Nike effectively turned this moment into a carefully orchestrated media campaign. Nevertheless, the dating of the letter several months before the release of the Air Jordan 1, the contention that Jordan actually did not wear the Air Jordan 1 when the NBA scolded him for violating its uniformity of uniform rule, and lack of confirmed evidence that the NBA banned any Nike basketball shoe or fined Jordan for wearing it does raise questions about the transgressive nature of Nike, Michael Jordan, and the Air Jordan 1. In a sporting world where a little bit of hype and hyperbole can go a very long way, Pete Forester was accurate when he wrote "the story is more of a fable than a description of reality."[15]

Nike continued to promote the Jordan brand, and "Money, it's gotta be the shoes" became a famous catchphrase uttered by Mars Blackmon in Nike's long-running Air Jordan commercials, which first aired in early 1989.[16] In the ads, Blackmon comically quizzes Jordan's athletic ability and demands Jordan agree that Nike Air Jordan shoes enabled him to perform amazing feats on the basketball court. The series of advertisements poked fun at the idea that athletic shoes could have anything to do with Jordan's ability on the court. In a sense, these advertisements reinforced the primacy and the uniqueness of the athletic body. Jordan represented the purest example of a gifted athletic body. The underlying joke was that Jordan would be a remarkable player regardless of what he wore on his feet. The commercials, though comedic, also reinforced the idea that shoes were a mundane component of the game. A quarter century later, the NBA returned to the shoe-banning business, and at issue this time was whether a shoe had the ability to enhance a player's performance. Surprisingly, the new banned shoe did not emanate from the technoscientific armories of the major athletic shoe manufacturers, but a small company aptly named Athletic Propulsion Labs (APL).

On October 19, 2010, the NBA issued a press release announcing that, effective immediately, the APL Concept 1 basketball shoe would be banned. This announcement caught the sporting world off guard because this shoe apparently was only the second shoe to be banned by the NBA in its history, but, more importantly, very few had ever heard of APL. The first shoe "banned" defined the era of basketball dominated by Michael Jordan. The APL shoe was an entirely different artifact; it had no association with a great player, or any NBA player for that matter. The NBA had long embraced shoe color and design creativity, but the APL shoe was the first to be banned because it created "an undue competitive advantage (e.g., to increase a player's vertical leap)."[17] What made the APL shoe, of which only a few thousand pairs had been purchased, worthy of a ban by the largest professional basketball league in the world? The simple answer is the potential of new technoscience entering professional basketball that could significantly direct the game away from narratives of individual athletic greatness.

APL contended that its Concept 1 shoe, released in March 2010, used the company's "proprietary Load 'N Launch Technology . . . to accomplish something never before achieved in the athletic footwear industry: a product that makes you jump higher instantly." APL designed the shoe to have "the Load 'N Launch device . . . implanted in a cavity in the forefoot of the shoe and [serve] as a 'launch pad' by taking the energy exerted by the player and increasing lift with the aid of an intricate, spring-based propulsion system."[18] It was a bold, if not revolutionary, assertion to claim that the company could provide athletes with a technoscientific device that would increase vertical leap. The patentees of this technology, John Bemis, Mark Goldston, Adam Goldston, and Ryan Goldston, submitted the initial patent application on May 18, 2009. Naming the patent "Forefoot Catapult for Athletic Shoes" probably did not endear the device to the NBA.[19] Although APL was a young company, it was not completely in the dark regarding the athletic shoe business. Mark Goldston previously promoted the "pump" technology seen in select Reebok shoes and was a marketing executive as well as the chief operating officer at L.A. Gear, a shoe company known for its popular brightly colored sneakers in the late 1980s and early 1990s.[20]

APL met with the NBA's executive vice president of basketball operations Stuart Jackson in July to gain league approval for the Concept 1.[21] On October 18, the NBA formally notified APL that the company's shoe would be banned because it violated section II.d of Rule NO. 2, which states: "All equipment used must be appropriate for basketball. Equipment that is unnatural and designed to increase a player's height or reach, or to gain an advantage, shall not be used."[22] The NBA concluded that APL's product gave an unfair advantage to all players using it. At first glance, it would seem that a ban by the NBA would precipitate financial crisis, but for APL, the ban produced the exact opposite effect. Soon after the NBA's announcement, the story went viral and the demand for the shoes spiked. In fact, the public desire and curiosity crashed APL's website, and it "took eight hours and eight new servers to get it back online."[23] Once back up and running, APL sold more shoes that day than it had in total prior to the announcement.

A significant part of the shoe's allure, priced at $300 a pair, was the NBA's ban. The ban implied that the shoes were amazing performance-enhancing weapons. For many amateur basketball players, APL presented an option for improving their game apparently so potent that the NBA deemed it worthy to oust from the game. The NBA's ban had no impact on recreational play, so those who could afford to experiment with these new shoes drove the demand. The shoes also fulfilled technophilic desires. A $300 technoscientific fix for what nature and training could not provide certainly was a bargain to those who purchased the shoes. But, outside of the fervor that the NBA's ban created, were there actual technoscientific grounds for banning the shoes, or did the NBA ban them to protect their brand by eliminating technoscience that could be seen as undermining the narrative primacy of the body and heroic play on the court?

The NBA may have had legitimate concerns because the shoes actually did work. In an unidentified study, APL concluded that "11 of the 12 participants jumped higher instantly in Athletic Propulsion Labs shoes compared to the leading brands [and] some participants saw an increase of up to 3.5 inches instantly in their vertical leap"[24] (figure 2.2). Most reports and reviews noted small to significant increases in leaping height.[25] Problematically for the NBA, and potentially for the sport of basketball, these results undermined the ethical work narrative that if a player wanted to improve his or her jumping ability, training was the only verifiable method to achieve this goal. And while there are many training regimens to help improve leaping ability, APL was suggesting that lacing up a pair of their shoes eliminated all such work. The ban eventually filtered its way down to high school basketball, where the National Federation of State High School Associations (NFHS) informed its members in early December that it too was banning the shoes.[26] Some state organizations, such as the Virginia High School League, banned the shoe as early as November 16, 2010.

Yet on February 3, 2011, Mary Struckhoff, NFHS assistant director and basketball rules editor and national interpreter, notified NFHS members that "based on discussions with Athletic Propulsion Labs, Inc., and on an examination of pertinent research, we have concluded that Concept 1 shoes do not violate Rule 3-5-4 . . . [therefore] the ban . . . should be immediately lifted."[27] Rule 3-5-4 stated that "equipment which is unnatural and/or designed to increase a player's height or vertical reach or to gain a competitive advantage shall not be permitted." This rule only varies from the NBA's rule on equipment in two words. Of course, this reversal raised the all-important question, why was the same technoscience that was banned by the NBA legal for players to use in high school? Evidently, the NFHS judged that these shoes did not violate Rule 3-5-4, so was this rule interpreted differently by the NFHS? Or was the implication that professional athletes have the potential to more effectively exploit APL's forefoot catapult? This inherently raised additional questions regarding what types of technoscientific equipment do not comply with the requirement that all equipment used must not "increase a player's height or reach, or to gain an advantage." If this was the requirement, many shoes other than the APL Concept 1 should have been examined.

For instance, on April 14, 2011, Adidas announced, with celebrated sneaker authority Bobbito Garcia, that it would release the first version of its adiZero Crazy Light basketball shoes. Foreshadowed the week prior on the Adidas Facebook page with the slogan "as light as _ _ _ _," at 9.8 ounces, the company touted it as the lightest basketball shoe ever produced. Roughly six months after the APL ban, Adidas did not shy away from boasting about how their technoscientific achievement would alter the game. The press release stated that "the revolutionary adiZero Crazy Light sets a new standard for basketball footwear at more than 15 percent lighter than the nearest competitor to help players be faster to improve performance on the court." Not only did Adidas assert that these shoes would give players an advantage, but the company also employed NBA All-Star Derrick Rose to support its claim. Rose contended, "Lighter footwear makes you faster and speed dominates. . . . adiZero Crazy Light is the lightest basketball shoe I've ever worn and it will no doubt make me faster on the court."[28]

Endorsement from celebrated sponsored athletes in the business of shoe sales is a familiar concept, but Lawrence Norman, Adidas vice president of global basketball, made even stronger claims than Rose. Norman declared, "No other basketball shoe compares to the new adiZero Crazy Light and it's a shoe that will revolutionize the game." He continued by outlining how "the ultra lightweight design is built for players who want to be faster, jump higher and perform better on the court. From the NBA's best to pick-up games across the world, players are asking for light footwear to help them become one step quicker—and the adiZero Crazy Light delivers this in a way like no shoe in history."[29] Arguably, the APL shoe's leaping boost could provide a comparable advantage to wearing a significantly lighter shoe over an eighty-plus-game regular season where players regularly run over two miles per game. Adidas did not appear to be the least bit concerned about a potential NBA ban of their shoe, which it proclaimed would reshape basketball by providing players with a technoscientific boost.

Herein lies the messy space for athletic shoes. It is difficult for publics to understand shoes as technoscientific performance aids because they are so familiar. Furthermore, the athletic superiority of NBA players has been so forcefully narrativized that it seems inconceivable that shoes could alter the way the game is played. Yet in athletic environments where differences in athletic ability are small, a slight advantage, like a lightweight shoe, could supply a substantive edge. Nevertheless, viewing publics are not particularly interested in entertaining this possibility because of its potential to rupture the heroic and culturally valued athletic narratives that NBA players represent. Currently, many sporting shoe companies are racing to develop and create even lighter shoes for all sports. In any sport that requires running, the idea of having less mass on an athlete's foot most certainly will increase performance. Though it can be argued that shoes do not substantively impact a game and that all shoes are relatively light, in sporting arenas, where every minute advantage matters, the weight of the shoes on an athlete's feet may contribute to the small differences between winning and losing. Studies on basketball players have shown that different positions require different footwear demands. Guards tend to favor lightweight shoes that allow them to capitalize on their speed and agility. Centers more often choose shoes that provide a high level of stability and will protect their ankles from injury.[30] Other studies have begun to examine if shoe aerodynamics can improve the speed at which athletes run.[31] Altogether, shoes, which traditionally have been thought of as basic and utilitarian, are much more than just protective foot coverings.

It's Gotta be the Ball

How one responds to the environmental conditions of a sport is a great determiner of an athlete's success or failure. Often the introduction of new equipment unsettles rote expectations of how a game should be played as well as reveals how some athletes are better than others at adapting to new technoscientific equipment. For many, sports such as basketball, football, cricket, and golf, the competitive task is to pass an object through or beyond a goal to score points. The way in which players score points differs depending on the sporting event, but all generally use some type of ball to do so. Though there are celebrated cases of "juiced" balls in sports such as American baseball, historically balls have received little critical attention because of their seemingly insignificant influence on play.

At the 2010 World Cup, however, the erratic behavior of the Jabulani soccer ball raised specific concerns about how a ball, and in this context all balls, can influence the game and how it is played. In its early use running up to the 2010 World Cup, goaltenders remarked that the Jabulani flew strangely, or "knuckled," when forcefully struck. Initially, many commentators dismissed these claims as unfounded complaints by coddled elite soccer players uncomfortable with something not perfectly the way they wanted it to be. Both Fédération Internationale de Football Association (FIFA) and Adidas aggressively defended the ball. But as time wore on, it became clearer that the ball did perform differently. Though players who did not put as much "English"—or spin—on the ball appeared to like the ball, some forwards, midfielders, and strikers began to complain that they were not able to curve the ball with traditional striking techniques.[32]

On closer examination, lab-based analyses found that the highly aerodynamic casing of the Jabulani ball made it a bit too slippery to "bite" into the air and curve like other soccer balls. Additionally, the ball's aerodynamic qualities produced unpredictable flight patterns above certain speeds.[33] So to an extent, the players were correct. But in thinking about technoscience and sport, what is interesting about this ball is that it maintained a level of its instrumentality because all competitors had to use the same ball. During the World Cup, all players used the Jabulani, so all competitors were subject to similar ball conditions. If the ball was a problem for one, it was a problem for all. There also was a growing sense that players just needed to figure it out. Anyone making their nation's World Cup team likely played the game since he was a small child, and potentially with far worse equipment. It was thought that great players would rise to the occasion and adjust. At the elite level of soccer, the ball is generally the same, but the ability to use different balls, such as in golf, can create a different set of technoscientific tensions.

In the 1970s, Fred Holstrom and Daniel Nepela designed the Polara golf ball to fly straighter when placed in the proper orientation. They designed this ball for casual players to have more success, and potentially more fun, but the new ball caused a stir within the staid game of golf. The Polara ball represents an instance in which a once-benign object of play became a source of cultural conflict about how to properly play the game. This ball, with its unique dimple pattern, came under attack from the USGA and the Golf Ball Manufacturers Association. These organizations claimed that by making play "easier," this technoscientific innovation undermined the history and traditions of the game.[34] The Polara and similar self-righting balls disappeared by the mid-1980s, only surviving in golf's underground. In 2006, the Polara ball reemerged to reignite decades-old debates about the golf ball and proper play. Governing organizations, as keepers of a sport's traditions through rules and regulations, deemed that this technoscientific device defied the cultural tradition of the sport of golf. That is, athletic bodies, not technoscience, should determine the outcome of sporting competitions.

Thus technoscience can be seen as not only altering a game but tinkering with the cultural values of sport. Though the sporting marketplace has grown in size and scale over the twentieth century, the cultural value of athletes and their performances has remained exceedingly high. It is not always a sport governing organization that determines if new technoscientific equipment undermines the game. In 2006, the NBA saw an organized protest against its efforts to introduce a new ball by a seemingly unorthodox opposition to new technoscience—the players.

On June 28, 2006, the NBA issued a press release beginning with the line "Technologically-advanced game ball has new design, better grip and consistency."[35] The NBA and Spalding—the ball's manufacturer and long-time NBA supplier—intended for the newly designed ball's "Cross Traxxion technology" to bring the game into a modern technoscientific age. This was a strange announcement for a traditional game known for its simplicity. A ball and a "hoop" are the basic requirements for play. The excitement of the game centers on how effectively, efficiently, and creatively a player can put a ball through the hoop. From an outsider's perspective, it is unclear what a newly designed and engineered ball could bring to the game. Since professionals and amateurs alike were not clamoring for a "better" ball, was this new ball a solution searching for a problem? The muted orange leather basketball, with its subtly nubby surface, had become a highly standardized artifact of the contemporary game. Part of becoming a proficient player involved being able to handle this "rock." Like many games in which a ball is the central component of play, a leather basketball had become deeply familiar to professional players, who dedicated countless hours to mastering the manipulation of the ball on the court.

In understanding the creation and deployment of a new basketball, it is important to understand the meaning of the ball within the game and how the institutions of basketball constructed it as a piece of instrumentalized equipment. But, it is equally valuable to discern how the introduction of a newly engineered synthetic ball revealed how central the historically and culturally rooted leather ball had become to the game, at least at the professional level. Dominant basketball narratives connected to skill and ability demanded that the ball not determine or influence the quality of play. It was one's commitment to the game that determined basketball success. Thus the design construction or material of the ball, in theory, should not matter in play. But as players used the new ball, it became abundantly clear that the leather ball correlated to success on the court, or at least in the minds of a significant portion of NBA players.

If any company possessed a knowledge of basketball history proficient enough to determine if the time was right for a new ball, it was Spalding. The connection between Spalding and basketball can be traced to just before the turn of the twentieth century, when at the request of the game's creator, James Naismith, A. G. Spalding and Bros. created a specialized leather-covered sphere with a laced-in bladder for the game in 1894. Over the next century, the ball changed very little. It lost its laces by 1937, and Spalding introduced a molded leather ball in 1949. With the market-wide creation of new materials for athletic use, Spalding developed a synthetic leather ball in 1972 and produced the ZX Microfiber Composite cover for the inaugural WNBA season in 1997. Though Spalding and other companies such as Nike expressed their technoscientific innovation through new basketball designs, the NBA continued to use Spalding's Horween leather basketball (it became the official game ball in 1983). That is, until the 2006–2007 season.

The NBA did not introduce the new ball quietly. It announced the ball on June 20, 2006, during a ceremony including NBA commissioner David Stern, Spalding's vice president Dan Touhey, NBA senior vice president of basketball operations Stu Jackson, Boston Celtic Paul Pierce, and NBA analyst Kenny Smith. The NBA revealed the black-cased ball just before the 2006 NBA draft at the NBA store on Fifth Avenue in New York City. Clearly, the choice of the location, timing, and cast of characters indicated that the NBA saw the new ball as a way to take a bold step into a new technoscientific future. The NBA highlighted the value and importance of the ball's technoscientific treatment by emphasizing that the ball was "a union of revolutionary design and breakthrough material."[36] By invoking the adjectives of "revolutionary" and "breakthrough," the NBA underscored the magnitude of the technoscientific disjuncture from the previous leather ball.

The NBA invoked technoscientific rhetoric as a means to distance the new ball from the old one. The juxtaposition was easy to make. The now "old" Spalding ball's exterior was made from leather tanned by the Horween Leather Company, a one-hundred-year-old company that is known for its traditional methods and prides itself on the fact that its "leathers are still made today, by hand, the same way they were generations ago."[37] These traditional practices produce some of the most coveted leathers made, but this handcrafting of natural materials was vastly different from the technoscientific foundation of the new ball. In differentiating the new ball from the old, the NBA emphasized how the design of the new ball diverged from the old. In terms of construction, the ball was "comprised of two interlocking, cross-shaped panels rather than the eight oblong panels found on traditional basketballs. As a result, there is more material coverage." Not only did Spalding and the NBA tout the new design, but they also accentuated how and why the new cover was superior. "The material is a microfiber composite with moisture management that provides superior grip and feel throughout the course of a game. Additionally, the new composite material eliminates the need for a break-in period, which is necessary for the current leather ball, and achieves consistency from ball to ball."[38] The NBA indicated that scientific methods, in the lab and on the court, had been used to fine-tune the ball's performance and design. The NBA noted that its Development League and several former NBA players, such as Steve Kerr and Mark Jackson, participated in its evaluation.

The NBA stressed standardization and grip as the key features of the new ball. In regard to standardization, the NBA quoted Dwyane Wade: "It will be great to get a ball that just feels the same no matter where you are and then it'll really give you that comfort that you need, either home or away."[39] Steve Kerr agreed: "I dribbled probably eight or ten different balls and they all felt exactly the same, which is really key as a player—you want to have the same feel dribbling the ball, and as a shooter, to have that grip, to be able to feel the grooves and your hands and finger tips on the ball and not have it slide off a little bit—that is really important."[40] As for feel and grip, Mark Jackson pointed out that it often took a month or more to break in a new ball, but he felt the break-in time for the synthetic ball was nearly nonexistent. Similarly, the NBA gathered quotes from legends George Gervin and Kiki Vandeweghe confirming that the new ball was easier to handle when it was wet from sweat. At the launch, everything appeared to go as planned, and the NBA proved it was a forward-thinking professional league excited about using the latest developments in material science to advance the game. Unfortunately for David Stern, the new ball quickly revealed itself to be more than a benign piece of technoscientific equipment. The ball became the central object of the early 2006–2007 season and raised fundamental questions about the applicability and necessity of new technoscience as well as the NBA's ability to effectively determine how and when game-altering technoscientific changes can and should be made.

Though many players tested the ball, it did not receive regular NBA play until the opening day of training camp on October 3, 2006. Shortly after the ball began to see rigorous play, complaints surfaced, ranging from its erratic performance to its stickiness. Dallas Mavericks owner Mark Cuban was so concerned about the ball that he enlisted University of Texas at Arlington physics professors Kaushik De and Jim Horwitz to study the ball. They marched the ball through a series of tests from October 14 through October 26 and made four primary conclusions when comparing the new ball to similarly inflated and broken-in leather basketballs. On examining the ball's bounce, they noted that "synthetic balls display measurably reduced return height than the leather balls—about 5% less on average, when dropped from about four feet." This potentially unfamiliar bounce performance could adversely impact the ways a player dribbled, passed, or rebounded during game play. When dry, the new ball displayed "a factor of two higher coefficient of static friction," resulting in a much more "grippy" ball. Yet when wet, the synthetic balls had "a coefficient of friction which is at least 30% smaller than similarly moistened leather balls," making them more slippery when wet. Finally, they established that the ball did not bounce as true as a leather ball. Their analysis determined that the synthetic ball deviated more than 30 percent when bounced. These numbers indicated that the synthetic ball was appreciably different in feel and play from a traditional leather basketball. This study, supported by one of the most outspoken owners in the league and right on the heels of the ball's introduction, concluded that NBA players were correct in sensing that something was amiss when they played with the new synthetic ball.[41]

Over the next few months, the NBA and Spalding went into damage-control mode in order to protect their public images and manage player concerns. The NBA's executive vice president of basketball operations Stu Jackson was responsible for interviewing players, and Spalding collected "its own samples and data from all the teams" to devise an appropriate solution.[42] Probably the most outspoken critic of the ball was Phoenix Suns All-Star point guard Steve Nash. Nash stated that he needed to keep his "hands from getting too dry because when they get dry, the ball just tears them apart. It's kind of like paper cuts."[43] As players had varying degrees of success adjusting to the new ball, the NBA held fast and reiterated to all concerned that the players would eventually familiarize themselves with the ball and embrace it. However, the National Basketball Players Association (NBPA) deemed this stance untenable, and on Friday, December 1, 2006, it filed a grievance with the National Labor Relations Board against the NBA for instituting the synthetic ball without proper player input and assessment.

The grievance also highlighted how the NBA and Spalding understood the ball to be an instrumental piece of equipment, whereas the NBPA considered the ball to be a fundamental component of a safe and productive work environment. In a sense, the NBPA demanded that the NBA embrace the ball as more than just infrastructural equipment of the game. The NBPA's complaint centered on the fact that Spalding and the NBA had not consulted with the union regarding the ball change, and that no active players participated in the ball's assessment. Three retired players—Mark Jackson, Reggie Miller, and Steve Kerr—were the primary evaluators with NBA experience, but they "spent less than an hour one day at Madison Square Garden shooting, passing and dribbling."[44] The only time the ball was used by active players was during the 2006 All-Star game, which was not considered to be an adequate evaluation period. Jerry Stackhouse, the Dallas Mavericks union representative, summarized player sentiment when he stated: "If it's something about the arenas and the fans and trying to enhance the game from a fan's perspective," the NBA should use its "expertise and business savvy to make unilateral decisions about that. . . .

When it comes to the actual game itself and when it comes to in between the lines, we should definitely have some input."[45]

In defense of the ball, Stu Jackson commented in October that the new ball brought the NBA into alignment with other professional basketball leagues around the world, most of which did not use leather balls. This was not the strongest rationale for switching away from the leather ball preferred by most NBA players in arguably the best professional basketball league in the world. NBA commissioner David Stern publicly acknowledged the complaints and reservations on December 5 when he stated: "I won't make a spirited defense with respect to the ball. . . . In hindsight, we could have done a better job."[46] In a very conciliatory tone, Stern declared: "Our players are unhappy with it, [and] we have to analyze to the nth degree the cause of their unhappiness. Everything is on the table. I'm not pleased, but I'm realistic. We've got to do the right thing here. And of course the right thing is to listen to our players. Whether it's a day late or not, we're dealing with this."[47]

After months of damning press and player irritability, on December 12 Stern and the NBA announced that the leather ball would return on January 1, 2007. Stern begrudgingly stated that though the "testing performed by Spalding and the NBA demonstrated that the new composite basketball was more consistent than leather and statistically there has been an improvement in shooting, scoring and ball-related turnovers, the most important statistic is the view of our players."[48] In the end, the "old," and potentially no less technoscientific, leather ball was returned to play with little fanfare. In this instance, it was not only the feel and sense of the seemingly older technoscience of the ball, but for the players there was also an emotional attachment to the ball. With fewer than five hundred players having the privilege to play in the NBA each year, it is a small group that has a strong sense of its place within its sporting culture. These players took the familiarity of the ball for granted until it disappeared, at which point they realized how consistent, trustworthy, reliable, and ultimately important the ball was to the game. In the end, the players reciprocated the ball's loyalty by demanding and fighting for the traditional leather ball to maintain its place within the game.

Games, players, and governing institutions tend to treat balls as functional or instrumental components of the game until something changes or goes awry. Historically, the manipulation of the ball is the primary ball concern, as in the cases of assumptions about baseballs being "juiced" or more recently the New England Patriots' use of underinflated footballs in the American Football Conference Championship Game against the Indianapolis Colts in January 2015.[49] The introduction of a synthetic leather ball by the NBA added a new wrinkle to ball narratives. Though the NBA and Spalding claimed that there was no appreciable difference between the leather and synthetic balls, game data from the season showed that players committed more turnovers with the synthetic ball and that turnovers decreased with the reintroduction of the leather ball.[50] As we champion the place of the body in competitive athletic endeavors, it is critical to remember that bodies train to play with and around the standardized equipment of games. When this equipment changes, athletes and their bodies must adapt to the power and effectiveness of game-altering technoscience.

It's Gotta be the Machine

Athletic gear consists of more than balls and shoes. It also includes mechanical devices that athletes use. These machines range from automobiles and motorcycles to tennis racquets and golf clubs. Similar to balls and shoes, narratives of athletic success designate machines as functional equipment that does not significantly influence or undermine an athlete's natural ability. In the world of cycling, for example, the dominant network of technoscientific machines is a bicycle. An athlete's ability to push his or her body to physical extremes determines cycling success. Historically, in the world of road cycling there are three markers of excellence. One is winning one of the three grand tours, Vuelta a España, Giro d'Italia, or Tour de France; the next is winning the world championship road race; and the last is setting a new hour record. The hour record, which challenges a cyclist to ride as far as possible in sixty minutes, has a long and storied history. Fans and competitors alike view this event as an objective way to evaluate pure cycling endurance, strength, and ability over time. Henri Desgrange, one of the original organizers of the Tour de France, set the first recognized hour record of 35.325 kilometers at Vélodrome Buffalo in Paris on May 11, 1893. The controlled and standardized environment of a velodrome makes an ideal location for this cycling effort. Competitors do not contend with the vagaries of pack cycling—there is no one to draft behind, no one to outsmart, and no period for rest. It is just a cyclist against time. The effort becomes a personal battle of the mind's ability to force the body to acquiesce to increasing levels of physical pain.

When Eddy Merckx announced his retirement from competitive cycling on May 18, 1978, he had won nearly every major cycling competition. By the end of 1971, his two Giro d'Italia, two world championships, and three Tour de France victories confirmed his status as one of the best cyclists of the twentieth century, but if he wanted to climb into the upper echelon of cycling's elite champions, he would have to ride the "hour." In the fall of 1972, after adding another Giro and Tour victory to his cycling palmarès, it became clear that it was time for Merckx to attempt the hour. At 8:56 a.m. on October 25, 1972, before the daily heat bathed Mexico City's velodrome, Eddy Merckx commenced his sixty minutes of suffering. Prior to his ride, most cycling journalists and fans believed that Merckx would put the record out of reach—and he did not disappoint.

Exactly one hour later, officials timing the event confirmed that Merckx had covered the unimaginable distance of 49.431 kilometers (30.715 miles)—farther than anyone had traveled on a bicycle in that short period of time. The distance of 49.431 kilometers in an hour was 0.778 kilometers farther than the previous record held by Ole Ritter. Minutes after completing his historic ride, a Mexican journalist asked Merckx if he thought he could ride farther and, if so, if he would try the attempt again. Merckx, one who was never shy about his abilities and accomplishments, stated that the mental strain, physical intensity, and unwavering concentration required for the hour-long ride was so immensely high that he did not think he could ride any farther and that he would never attempt the record again.[51] He held true to his word. That morning in Mexico City was his only attempt. His unremitting reiterations to never ride the hour record again shaped public perception of his record in relation to all attempts before and after. If "the Cannibal"—Merckx's nickname, describing his relentless cycling speed, power, and aggression—said that 49.431 kilometers was the limit of his ability, then this must be the limit of all physical ability. Merckx's record eventually fell in the winter of 1984 to Francesco Moser, arguably not because Moser was a superior athlete but due to the aerodynamic advantages of Moser's bicycle. In the decades following Merckx's hour ride, technoscientific innovation in cycling spurred a series of riders to make this seemingly unsurpassable mark a quaint historical footnote and turn professional cycling into a spectacle that was as much about athletic struggle as it was about material science and engineering innovation.

In an attempt to reclaim cycling from the onslaught of technoscientific ingenuity in the decades following Merckx's record-breaking ride, the Union Cycliste Internationale, cycling's governing body, subsequently turned back to Merckx and the hour record as a means to again elevate the body over the machine. The UCI went to work and passed a series of legislative actions to aggressively push back against the technoscientific upheaval. On October 8, 1996, the UCI's Management Committee issued the Lugano Charter. This governing document aimed to curtail "the potential danger and problems posed by a loss of control over the technical aspect of cycling [and] recall that the real meaning of cycle sport is to bring riders together to compete on an equal footing and thereby decide which of them is physically the best." The Lugano Charter continued with its anti-technoscientific tone and argued: "If we forget that the technology used is subordinate to the project itself, and not the reverse, we cross the line beyond which technology takes hold of the system and seeks to impose its own logic."[52] It is clear that the UCI perceived technoscientific inventiveness as a troubling wedge whose outcomes could undermine the history and traditions of cycling based on physical feats of otherworldly performance.

In line with the Lugano Charter, the UCI addressed minimum bicycle weight on June 11, 1999. The concern about weight stemmed from increasingly lighter bicycles potentially influencing the outcomes of races, in particular those traversing mountainous terrain. In this release, the UCI was unambiguous about the reason for its decision. "The International Cycling Union (UCI) wishes to specify that, in its concern to protect equal chances and the primacy of man over machine in cycling races, its Management Committee has decided to limit the minimum weight of bicycles in competitions (road, track and cyclo-cross) to 6.800 kg from 1st January 2000."[53] For many, this decree was an outrage. To deny professional cyclists the best equipment available seemed decidedly regressive. Furthermore, this mandate did not align with the realities of the consumer bicycle marketplace because it was and still is quite easy for anyone with the financial means to walk into a bicycle shop anywhere around the world and purchase a bicycle that weighs less than 6.8 kilograms. Finally, on September 9, 2000, the Management Committee issued the following statement regarding the hour record:

> With the aim of re-launching this very appealing discipline, whose history is tied up with the performances of the greatest cycling champions, reforms are needed based on the modifications made to the regulations which have been applied since 1997 (Lugano Charter). In view of the fact that the new cycle sport regulations would make the current record virtually impossible to beat, the Management Committee has therefore decided to create a UCI Hour Record. . . . From today, the UCI Hour Record is the one that Eddy Merckx achieved in Mexico on 25th October 1972, covering a distance of 49.43195 km. This UCI Hour Record can only be attempted if the equipment is presented and checked beforehand by the UCI and it must be similar to that used by Merckx. . . . This distinction will allow the respect of a long tradition of a classic cycling speciality, without endangering the vital modern aspect of our sport.[54]

This final declaration returned cycling to the valorized representations of Merckx on that fateful day in 1972. But this homecoming of sorts was not as technoscientifically pure as the UCI wanted cycling publics to believe.

The Merckx bicycle, once thought of as a minor player in his record-setting hour ride, was an equally fine-tuned member of the hour record team. Bicycle racing technology into the 1960s was much more about art, tradition, craft knowledge, and skill than it was about science, technology, and engineering. But Merckx leveraged his collaboration with frame-builder Ernesto Colnago to capitalize on a new technoscientific way of thinking. Colnago, from his small workshop in Cambiago, Italy, began reexamining bicycle fabrication in the 1950s, and by the 1960s he designed and built some of the most innovative and sought-after racing bicycles. Colnago built his first bicycle for Merckx in October of 1970 and regularly built at least twenty bicycles each racing season for Merckx. Colnago's experimentation with new designs, materials, and techniques fit well with Merckx's deep interest in technology and lightness. In reflecting on his relationship with Merckx, Colnago commented: "No detail was too small for Merckx to pay attention to. I remember that one time, I told him I could file a lug a certain way that might save a few grams and he was excited; 'Do it that way!' . . . We innovated a lot of new things for Eddy; titanium parts, drilled chain, super lightweight bicycles and parts. We made rims that were 260 grams for him. . . . We looked at every detail to save weight and do special things for Merckx and that helped us improve our total approach to bike building."[55]

Due to their successful relationship building winning road racing bicycles, Merckx trusted Colnago to build a cutting-edge bicycle for his hour record attempt. Merckx, who eventually spent $20,000 of his own money in pursuit of the hour record, approached the bicycle design with focus and precision.[56] He assembled a team to plan the hour record challenge, including everything from the ride location to the bicycle design, and drew on the knowledge of Giorgio Albani (former Italian road racing champion and the directeur sportif of Merckx's Molteni road racing team), Guido Costa (Italian national track cycling coach), and, of course, Colnago. Colnago recounted that their preliminary meetings were not as successful as he had hoped. While they all agreed that the attempt should take place in Mexico City rather than in a velodrome in Europe because the thinner air produced less wind resistance, Costa derailed the bike design discussion by arguing for a design similar to those made for Olympic-length track cycling events.

Colnago fundamentally disagreed. He wanted a more robust bicycle because Olympic track events lasted four or five kilometers instead of the roughly fifty Merckx aimed to ride. Colnago's point of view eventually won out. He took over the bicycle design and concentrated his efforts on building the lightest machine that could withstand the rigors of the hour ride. He estimated that it would take Merckx approximately one kilometer to reach the target speed of 50 kmh. Once up to speed, the stress on the bicycle decreased substantially. From these insights, Colnago produced two handmade bicycles, each weighing a stunningly svelte 5.5 kilograms (12.125 lbs.). Comparatively, a professional road-racing bicycle weighed approximately 11 kilograms at the time.[57] Colnago remembered the bikes and his aggressive innovations fondly:

> [It was] hard to imagine how much work we put into building those
> two bicycles. . . . In those days, the tubing was 6/10 to 4/10 wall
> thickness. A guy like Merckx, who weighed 72kg, 1m 85cm high with

incredible power, well, to build a superlight bicycle for someone like
that took a lot of courage. I lightened everything; the cranks, drilled
out the chain, because we wanted the lightest material and well, you
couldn't buy a Regina Extra chain with holes drilled in it! . . . Do you
realize that the tubular tires we used, the Clement number 1 Pista,
weighed 90 gr. for the front and 110 gr. for the rear! No one else could
have put this bike together. For example, no one in Italy could weld
titanium, so I went to Pino Morroni in Detroit, America to weld the
special lightweight titanium stem for Merckx's Hour Record bicycle.[58]

The only thing they were unable to source in Mexico City was he-
lium for the tires. During the record attempt, Colnago feared that the bike
would fail. No one with so much physical power had attempted the hour
record on such a light and drilled-out bicycle. At about the halfway mark,
Colnago, who had crisscrossed the velodrome with the second bike just in
case something went awry, felt comfortable enough to have a seat and enjoy
Merckx at his best.

In the following days, newspapers in Mexico, France, Italy, and Belgium gave detailed accounts of Merckx's success, whereas American newspapers such as the *New York Times* printed very short news briefs.[59] The
paucity of images associated with the reporting of this event is interesting.
Most of the photographs chosen were headshots of an exhausted Merckx.
Very few photos show Merckx actually riding. (Arguably, Merckx on a bicycle was a familiar sight for an early 1970s cycling enthusiast.) Recounting
the effort required to break the hour record resonated more with readers
than the actual process by which Merckx broke the record.

Merckx's heavily reproduced headshots took on new meaning after
2000, when the UCI, in its effort to return cycling to its "roots" and champion human performance over technoscientific innovation with the Lugano Charter, rewrote its record books and wiped away all the hour records
that came after Merckx. Merckx, his ride, and his bicycle quickly migrated
from a historically valuable record-breaking moment to representing a
fundamental shift in the way the professional sport of cycling would use
and interpret technoscience and athletic bodies.

After the Lugano Charter went into effect and changed the rules and
regulations for subsequent hour records, more images of Merckx actually

cycling during the hour ride began to circulate. This may have been due in part to the easy access to images in a digital age, but these images represented literal and metaphorical snapshots of the new UCI-mandated reference point. This new referent was devoid of CAD-designed bicycle frames, carbon fiber composite components, and performance-enhancing pharmaceuticals. In 2000, the UCI ruling redefined these images and Merckx himself as the new, yet old, standard of pure human cycling performance and achievement, which was in stark contrast to a technoscientifically derived machine. The UCI also deployed Merckx to curtail the public flogging the drug-riddled sport received in the press during the last five years of the twentieth century. Thus it comes as no surprise that the UCI, in an effort to maintain its brand, returned to its most valuable and trustworthy living legend, Eddy Merckx, and to a moment of "pure" cycling performance and achievement, when people believed and trusted that the man was truly greater than the machine. The UCI, and its return to Merckx and his hour record in 2000, was at the leading edge of emerging technoscientific apprehension expressed by sport governing bodies.

Seeing Merckx as the complete and perfect cycling specimen demands that we ignore and overlook the sophistication of the machine and the potential administration of performance-enhancing agents. It is irrefutable that Merckx's machine was as technoscientifically advanced as any bicycle could be in 1972, and it undoubtedly gave him an advantage. Furthermore, it is hard to believe that Merckx, who tested positive for banned substances three times during his career, rode only fortified by a breakfast of toast, ham, coffee, and cheese brought with him from Belgium. Images of his Mexico City ride are at such a distance that the eye can neither determine the adjustments and amendments to the machine nor adjudicate a pharmaceutically augmented body by sight. Compared to wind-cheating disc wheels, tear-shaped bicycle frames, and aerodynamic helmets that increasingly came into use after Merckx's hour record, Merckx's bicycle looked "normal." His body was considered special, unique, and decidedly abnormal, in that he did not need any more than breakfast to put the hour record out of reach. But this mythology, like most technoscientific and sporting mythologies, serves a social and cultural purpose. The mythological Merckx, as the perfect synergy of common bicycle and uncommon body, maintains the illusion that an athletic body will always be vastly more important than any device in the final outcome of a sporting event.

Swift changes in rules and controversies precipitated by the implementation of new statutes are not new within sport. Yearly, some device, practice, or procedure is deemed problematic by a national or international governing body. Recently, judgments regarding the improper or unsportsmanlike use of technoscience have pushed viewing publics and even governing bodies to discern whether the offending technoscience is an alteration to a body technique (i.e., the ways in which athletes use and manipulate their bodies), to the physical body itself, or to the artifacts of the game. The public and institutional queries into cyclist Fabian Cancellara's successes in the spring of 2010 bring these issues into higher relief. By winning cycling's 2010 spring classic, Paris-Roubaix, on a traditionally cool and gray spring Sunday in northern France, Fabian Cancellara completed the "double." His victory in the Belgian classic, the Ronde van Vlaanderen (aka the Tour of Flanders), the weekend prior made him only the tenth rider in cycling history to win both races in the same season.

These victories, among others, confirmed Cancellara's place as one of the best cyclists of his generation. Known for power and strength, his list of achievements includes multiple individual time trial and road race victories, junior and senior world championships, and bronze and gold Olympic medals. In the weeks following Paris-Roubaix and the Tour of Flanders, a story with a technoscientific twist surfaced that would question Cancellara's victories. In the context of cycling's recent history, replete with instances of competitors using or attempting to use performance-enhancing substances, one would assume that critics of his winning rides contended that illegal or banned substances helped to drive Cancellara to victory. Yet it was not his body or enhancements to his body that were of concern. It was his bicycle and rumors that an electric assist motor powered his performances.

On May 29, 2010, an Italian video taken from an RAI report surfaced on YouTube describing how an electric motor could easily be hidden within a bicycle frame.[60] The video went viral, demanding supporters and detractors weigh in.[61] It featured former professional cyclist Davide Cassani operating a motor hidden within the seat tube of a racing bicycle. During the demonstration, Cassani proclaimed matter-of-factly, "I could win a stage in the Giro with this bike."[62] Though he potentially made the comment in jest, Cassani's position as a former elite professional and cycling commentator on Italian television lent a high level of credence to his remarks that this device was so potent that even an out-of-shape former professional cyclist could compete and win at an elite cycling level. After Cassani's demonstration, the clip centers on two sections of televised race footage from Paris-Roubaix and the Tour of Flanders. The piece from the Tour of Flanders focuses on Cancellara's hand movements and the ease at which he rode up the slippery and excessively steep race-defining cobblestoned climbs on the parcours—the implication being, of course, that Cancellara's display of power and strength was above and beyond the ability of contemporary professional cyclists.

For Paris-Roubaix, the video concentrates on another suspect hand movement and an apparently improbable acceleration that launched Cancellara to victory. Once again, the video (slowed down for greater emphasis) implies that Cancellara's performances were beyond what was reasonable within elite-level professional cycling. Thus if he was not doping (he passed drug tests as the winner of both events), the only reasonable explanation for his otherworldly show of power was that he must be using some sort of illicit mechanical assistance hidden from other competitors, the observing publics, and cycling's governing body. Initially, Cancellara laughed off the reports of his motorized success. But as the wave of questions into his performance swelled, he responded quite forcefully. He did not mince words when he said, "It's so stupid I'm speechless. . . . I've never had batteries on my bike."[63] But in a sport where finding, building, and sustaining appreciable competitive advantages (legally or illegally) has been central to its competitive history, these seemingly extraordinary claims of motorized doping gained traction.

In May 2010, two Italian newspapers, *L'Avvenire* and *Il Giornale*, published articles indicating that the concept of this type of device had been circulating since late 2009. The UCI's initial response was to state that this technology was not a problem within cycling. Enrico Carpani of the UCI declared: "We do not have any knowledge if this product is already in use in competitive cycling. . . . At this point in time, we don't have any evidence that leads us to the conclusion that this kind of engine is already in use in the peloton. But our equipment commission will follow this issue very carefully because they are obviously interested in everything that could affect cycling in the future. The UCI is studying the machine to find a method to detect it."[64] Since these reports surfaced during the Giro d'Italia in May, the race directors also found it necessary to address the issue. Assistant Race Director Stefano Allocchio told Italian news agency *ANSA* that "there are no souped-up bikes at the Giro. . . . According to all the checks that have been done, all the bikes are ok. The chief judge is very attentive—if there was something unusual, he would have seen it straight away. . . . I understand it's something the UCI ha[s] been looking at since last November but at an amateur level, not a professional level. Everything is okay here at the Giro."[65]

In the *Il Giornale* story, Marco Bognetti—previously a member of the material commission and consultant to Jean Wauthier, the current head of the materials unit at the UCI—told a different story. He indicated that "it's all true, there's a suspicion that there are teams and riders who used a 'pedal-assisted' bike. . . . We were first told about it last July, during the Tour de France. We first heard about it from the USA and it set alarm bells ringing. . . . We've discovered that it could save a rider between 60 and 100 watts, which is an enormous advantage in the finale of a race." Bognetti confirmed that the UCI was actively engaging in building preventative measures. "Our technicians are working on a special scanner that will discover the hidden motors inside the frames. All the bikes at the major races will soon be checked."[66] Shortly thereafter, on June 18, 2010, the UCI confirmed that it would scan bicycles for the then-upcoming Tour de France, and by 2012 the UCI had a verifying labeling system in place.[67]

A device named the Gruber Assist became the chosen suspect for the machine in question. The Austrian manufacturer, Gruber Antrieb GmbH and Co., produced a seat-tube-installable motor to assist in the powering of a bicycle through the bottom bracket and the crank arms. When the story first broke, interested news organizations bombarded Gruber with inquiries about its device. This interest proved to be great marketing for Gruber, but it also implied that the company may have been complicit in this illegal behavior. In explaining the application of this form of assistive technoscience, the term *mechanical doping* gained traction.[68] From a technical standpoint, Gruber was let off the hook pretty quickly. First, its product could not fit inside the bicycle that Cancellara rode. The Gruber Assist requires a seat tube with a diameter of 31.6 mm, and the Specialized Tarmac SL3 bicycle Cancellara rode had a seat tube diameter of 27.2 mm.[69] Though Gruber faded from inquisition, it did not quell concerns that smart engineers could alter the device or the bicycle in order to insert the device into a carbon fiber monocoque bicycle frame.

Specialized created its bicycles from sheets of carbon fiber specifically layered in a single mold and heated to high temperatures, which made building a Gruber Assist type of device into a rideable bicycle nearly impossible. Furthermore, it appeared that the device was only capable of providing power assist between thirty and ninety revolutions per minute (rpm). Elite-level cyclists regularly ride faster than 90 rpm, thus its practical use was highly unlikely. Gruber's own Julia Timmerer confirmed these drawbacks.[70] Nevertheless, the concerns did not dissipate. In January 2015, the UCI added a new provision to its cycling regulations to warrant against "technological fraud."[71] Riders committing technological fraud would be disqualified from the race at which they were caught, suspended for a minimum of six months, and issued a fine between 20,000 and 200,000 Swiss francs. Teams involved in technological fraud would receive a suspension of at least six months and a fine between 100,000 and 1 million Swiss francs. Though the UCI performed an increasing number of systematic bicycle checks at major international races, it was not until January 30, 2016, that it first found a motor in a competition. Nineteen-year-old Belgian cyclist Femke Van den Driessche was caught with the motor during the 2016 women's under-23 Cyclo-cross World Championship race. This discovery finally confirmed the years of rumors and speculation.[72]

What is interesting about this narrative turn in technoscience and sport is that it confirmed the collective societal fear and belief that athletes will stop at nothing to gain a competitive advantage and that we live in a historical moment where current technoscience can and will influence the outcome of sporting events. Within the historical backdrop of professional sports, an athlete "cheating" is not surprising at all. It is partially due to this history that a story of a cyclist using a bicycle with a motor in it gained international attention. Cycling publics have come to suspect athletes, but the twist in the story is clearly the implicated technoscience. The idea that an elite-level cyclist or team would use an electric motor to assist their competitive efforts was completely new but well within the sphere of everyday technoscientific knowledge.

The device was not some futuristic technoscience contrivance; rather, it was an easily readable and recognizable electromechanical motor. Yet this device fell outside of generally understood technoscientific augmentations within sport. It was not a change in body technique. It was not external, meaning that the suspect machine was not visibly affixed to a cyclist's bicycle or body. The subsequent assumption was that the technology was internal. This wrinkle makes it difficult to read and interpret for sporting publics because cycling's internal technoscientific enhancements are mostly confined to bodies, not to the technoscientific machinery of the sport. "Invisible" pharmaceuticals circulating within bodies have been well understood, but unseen illegal mechanical devices surreptitiously hidden within a bicycle frame had not been conceptualized by the cycling public. Since there was little familiarity with this new technocultural frame, two older and more familiar words merged to form the new term *mechanical doping.*

Mechanical doping has multiple inflections that inform contemporary and historical concerns about the future of sport and technoscience. Journalists deployed the term to easily explain the invisible nature of the advantage that an electromechanical device could provide. In this case, the use of the word *doping* is not arbitrary. When one thinks about doping in cycling, images of blood-boosting pharmaceuticals are often the first that come to mind, not battery-powered machines that add auxiliary power to a competitor's pedal stroke. The very idea that these types of devices exist opens a new doorway into undermining cherished beliefs in human versus human sporting competitions and is troubling to many. In many ways, mechanical doping can be viewed as much worse than pharmaceutical doping. If an athlete injects his or her body with a performance-enhancing substance, the body still has to do work. Each body is unique and reacts differently to the substances it consumes. The organic complexity of the body often makes pharmaceutical doping part medical technoscience and part art. An electromechanical device has much less variability than a performance-enhancing substance. The body does not need to integrate the mechanical device; it only needs to react to the power received from the machine. In a sense, this may be the ultimate form of doping.

What is fascinating about emerging understandings of how material artifacts of sport influence outcomes is how we currently do not have a unique language to explain this reality, so the language of performance-enhancing drugs has co-opted rhetorics around the objectionable roles that new and emerging technoscience plays within sport. The predominance of performance-enhancing drugs in the modern sporting era has been defined by technoscientific work that has made its way to the playing field. Since sport governing bodies have elevated the use of performance-enhancing substances to the most evil form of cheating in competition, technological or mechanical doping has become a familiar shorthand that resonates with all concerned. Thus the linguistic power of performance-enhancing drugs has reached the point where electromechanical devices such as the one used by Femke Van den Driessche are lumped into doping. This, of course, is not new. Similar language has been applied to "juiced" baseballs and other forms of sporting enhancement.[73] Though the implication of performance-enhancing drugs is valuable, it is equally important to pull interpretations of technoscientific devices away from the pharmaceutical-drenched world of doping to understand how and why technoscientific alterations to sport seen in the vehicles competitors drive, the balls they play with, the devices they hit balls with, and the gear they wear while competing do not receive substantive analytical attention.

In each of the cases discussed, sporting cultures made sure they suppressed a potentially transformative technoscientific innovation. For athletic footwear, most sporting cultures actively ignore the ways in which athletic shoes aid performance. From traction to comfort, shoes have always changed the game. Whether the NBA banned the APL Concept 1 for performance reasons or to protect its relationships with major athletic footwear brands, it promotes and sells collective and individual feats of athletic ability, and the shoe as a performance enhancer works against this effort. Interestingly, the NBA's choice to introduce a new ball unraveled when the players decided that this technoscientific alteration fundamentally changed the game. In a similar effort to repress technoscience in the world of cycling, the UCI chose to make specific legislative changes and situate them within a valorized narrative of cycling. By linking their rule changes to an iconic moment and individual, the UCI skillfully constructed a historical rationale for its decision to limit technoscience.

Conceivably for these sporting cultures, more was at stake than cherished records, histories, and traditions. Each sport reacted as if the sport itself was under siege from a foreign and potentially threatening technoscientific invasion. This fear of technoscience overtaking a sport is clearly reflected in the broader society's contemporary fears of a future in which machines control, take over, and eventually eliminate humanity. Films from *2001: A Space Odyssey* to the *Terminator* and *Transformer* series cinematically represent this angst. Similar to these films, the feared technoscience in sport does not come from some external force but is man-made, and thus the pain and suffering is self-inflicted. This is the place where contemporary sport exists. The sporting, cultural, and economic demand for upward trajectories in all forms of athletic performance necessitates the use of artifacts produced in our increasingly technoscientifically driven world. Unfortunately, sporting cultures cannot blindly hold onto historicized traditions while simultaneously expecting increases in athletic performance and ability. We have invested so deeply in the heroic narratives of athletic prowess that it leaves very little space to understand and assess the roles of technoscience in augmenting and influencing the gear and equipment of the games we love. In a world where trust in sport has been honed narrow-thin, paranoia and the fear that something or someone is doing something illegal is easy to create. The examples of athletic shoes, a new basketball, and cycling legislation around equipment show how much effort will be spent to reclaim sport as a human endeavor when technoscience is no longer a basic component of sport.

3

Disabled, Superabled, or Normal?

Oscar Pistorius and Physical Augmentation

When South African sprinter Oscar Pistorius stepped onto the track at the Olympic Park stadium in the East London district of Stratford to run in a 400 m qualifier on the morning of August 4, 2012, he made history—again.[1] He became the first visibly disabled athlete to compete in a track event during the modern Olympic era.[2] Pistorius made it all the way to the semifinals, and by all accounts this was a success on multiple levels. His last place semifinal finish precluded him from competing in the 400 m final, but the mere fact that he competed in the London Olympic Games proved to be a turning point in contemporary sporting competition and our understanding of how technoscience destabilizes the exceedingly imprecise bodily categories of normal, less-abled, disabled, or superabled. Pistorius also broke the unspoken cultural agreement about the sporting segregation of able and less-abled bodies. His eventual participation in the able-bodied Olympics directly contested historicized forms of compulsory able-bodiedness.[3] He did not diminish or mask his prostheses, nor did he attempt to pass as more able-bodied. Instead, Pistorius embraced, accentuated, and potentially exploited the hybrid nature of his body, forcing sporting cultures to viscerally deal with athletic potentialities and realities when competitions are no longer differentiated solely on the murky distinctions between those who are able-bodied and those who are not (figure 3.1).

Figure 3.1. Oscar Pistorius and his carbon fiber composite prosthetic limbs; Olympic Summer Games, London, 2012. He pushed sport to reconsider the boundaries between the human body and technoscientific equipment. PCN Photography / Alamy Stock Photo

The participation of Pistorius, or someone like him, in the Olympic Games had been imminent for some time. It was a moment that was welcomed by some and feared by others, and also can be interpreted as a cultural breakthrough in sport—a fundamental shift in how global society understands less-abled bodies as well as accepts the technoscience that allows these bodies to freely live and compete. Less-abled bodies and the devices allowing them to function more easily have been accepted readily within most walks of life; however, one place these technoscientific advancements have been questioned heavily is within the context of sport. The overwhelming concern is that these devices, instruments, or practices will not just allow less-abled athletes to have a fair or equal opportunity to compete but will give these individuals an undeserved and unearned advantage. It was this historically, socially, and culturally rooted technoscientific fear that made Oscar Pistorius's runs so important.

When couched in the sporting mantra of fairness, vocalizations of technoscientific fear can be disturbing. When Olympic sprinting legend Michael Johnson questioned Pistorius's ability to compete, he presented himself as someone who was not attacking Pistorius as an athlete or his carbon fiber prostheses but, rather, as someone who was interested in protecting track and field from a competitor with an unknown, undertested, and potentially unfair advantage. Johnson stated, "My position is that because we don't know for sure whether he gets an advantage from the prostheses that he wears it is unfair to the able-bodied competitors."[4] By contending that the evidence regarding whether Pistorius's prostheses gave him an unfair advantage was inconclusive, Johnson positioned himself not only as a guardian of the purity and traditions of track and field but also as a defender of all sport from the incursion of new and seemingly event-altering technoscience. From Johnson's perspective, his comments are not specifically directed at Pistorius but are geared toward building boundaries from which to maintain the physical authenticity of sport. Unfortunately, the irony of these and similar contentions gets lost in the crossfire of the debate.

These damning statements would be infinitely more meaningful if none of the athletes Pistorius competed against—or no athlete in the past—had used any sort of technoscientific tool, instrument, or practice, but this is not the case. For instance, Australian Cathy Freeman won the women's 400 m gold medal at the 2000 Sydney Olympic Games. The racial narrative power of a woman of Aboriginal descent winning an Olympic medal in Australia was equal to the technoscientific narrative significance of Freeman winning the final as the only competitor wearing a hooded bodysuit designed by Nike. The irony is that Johnson also won a 400 m gold in Sydney, and as a Nike-sponsored athlete he would certainly have known about the suit. Arguably, Freeman wore the suit for its aerodynamic advantages and not as an opportunity to garner more fabric landscape for her Australian colors. Clearly, all the athletes competing in the 400 m races at the 2000 Olympic Games wore aerodynamic, country-colored outfits and the "fastest" technoscientifically designed and engineered shoes available for the competition. These familiar technoscientific devices pale in comparison to those out-of-sight technoscientific instruments that athletes use at training facilities around the world.

Thus arguments implying that Pistorius's prosthetic limbs created an unfair, or at best an unclear, advantage disavowed all the other technoscientific artifacts, practices, and knowledge that go into the production of elite-level sprinters and athletes in general. The metafear is that we have reached a point within sport where technoscience can trump genetics and talent, perseverance and motivation, and training and effort. As a result, the use of game-altering technoscience is perceived as "unnatural." What is natural and unnatural has been and will be endlessly debated, but this specific case provides an opportunity to think about what is at stake for sport if it continues to base one of the barriers between who can and cannot compete on the continually moving targets of natural and unnatural.

Pistorius's presence at an elite level of sport raised fundamental questions about what is considered a legal use of technoscience. Simply put, the boundaries between illegality and legality depend on how technoscience is perceived to provide an athlete with aid beyond what is deemed as "normal." Each sporting culture continually renegotiates balance between technoscience and the distinctions between normality and abnormality. *Normal* is a relative term, and scholarly fields such as disability studies have problematized the language and contexts of normality and abnormality.[5] In the worlds of sport, *normal* can be a euphemism deployed to delineate a boundary between fair and unequal competition. In this sense, technoscience that gives an athlete a significant advantage above and beyond the level of functional equipment on a field of play is viewed as unfair and will most likely be seen as illegal and subsequently banned.

But Pistorius is unique in that he understood his body, like most people who have used prostheses from a very early age, to be profoundly normal. This perspective has been welcomed in larger society, but it became an issue when he decided that he wanted to compete against non-prosthesis-using athletes in the able-bodied Olympic Games. The ethical and moral issues raised by his effort to compete provide insight into apprehensions regarding the use of technoscience in sport. His body and his prostheses provide an opportunity to explore the fundamental differences in the ways in which sporting cultures conceptualize bodily repair, assistance, and augmentation.

Most sporting cultures, overall, consider the repair of an injured or damaged body to be an acceptable form of medical treatment. Sports medicine exists to effectively and efficiently repair bodies and return these bodies to the field of play as quickly as possible. For instance, the game of baseball widely accepts ulnar collateral ligament (UCL) reconstruction surgery, which repairs the ligament connecting the humerus and the ulna, therefore stabilizing the elbow, as a justifiable and welcomed treatment to an injured athlete's arm. Frank Jobe first performed this surgery on Los Angeles Dodgers pitcher Tommy John, who damaged his UCL on July 17, 1974. Jobe replaced the damaged ligament with a tendon from John's opposite wrist. For the procedure, Jobe "drilled holes in the ulna and humerus and threaded the graft from John's opposite wrist through them in a figure-8 pattern" in order to repair the damage.[6] By all estimations, John had an extremely slim chance of returning to professional baseball. But after eighteen months of rehabilitation, he miraculously returned to pitch and played until 1989. Currently, nearly 15 percent of pitchers in Major League Baseball have had the procedure, colloquially called Tommy John surgery, and "92 percent of elite pitchers with reconstructed UCLs return to their prior level of competition for at least a year."[7]

This groundbreaking surgical procedure entirely changed the outlook on this specific injury and innovatively created a new way to save a pitcher's career, but it may soon morph into a procedure acceptable for pitchers regardless of age or injury. Baseball observers such as Lindsay Berra have raised questions about the procedure's use and potential abuse by increasingly younger players. Instead of fixing bad throwing mechanics, pitchers may rely on this surgery to backstop their impending injuries. The surgery's success allows athletes to abuse their bodies with a high guarantee that the damage can be repaired. Thus is this type of surgery simply repairing the body and bringing it back to a natural state, or is it augmenting a broken and abused body? This and similar types of medical procedures require us to collectively think about what it means for the future of sport if we fully embrace a new era of injury repair in which material science, instead of just simply repairing damaged tissue with bio-materials, can "guide and stimulate the innate healing response of the body" to push the future evolutions of athletic ability.[8]

Assistive technoscience comes in many different forms and causes varying degrees of concern. Some forms of technoscientific assistance, such as dietary supplements and scientific training, are widely accepted. But using visible tools that assist the body during play, beyond basic support such as a knee brace, can be called into question. Beyond basic assistance, technoscience can be seen as providing a competitor an unfair advantage. Casey Martin and the use of golf carts were at the center of a pathbreaking lawsuit regarding assistance at the turn of the twenty-first century. Martin, who was born with Klippel-Trénaunay-Weber syndrome—a congenital circulatory condition that impairs the circulation of blood back to the heart—sued the PGA on the grounds that the Americans with Disabilities Act (ADA) allowed him to use a golf cart during tournament play. The case made it all the way to the United States Supreme Court.[9]

The PGA brought out some of its most important legends, including Arnold Palmer and Jack Nicklaus, to testify on its behalf that a major part of the sport of golf is walking and that using a golf cart provided an unfair advantage. The court ruled in favor of Martin. The 7-2 majority decision stated that the PGA could not prove that walking the golf course was "an essential attribute of the game itself."[10] The PGA contended that the act of walking inserted "the element of fatigue into the skill of shot making," but the court determined that the goal of the game is to put the ball in the hole and that unpredictable conditions such as weather, terrain, or chance more profoundly influenced the game than the physical stress of walking.[11] Antonin Scalia, joined by Clarence Thomas, wrote the dissenting position and questioned whether the ADA should be used to alter the history, tradition, and rules of golf or any game. Those against the decision opined that it would open a floodgate of cart-driving professional golfers, but this has not been the case.[12]

Conceivably, governing bodies as well as invested fans want to determine if an athlete uses an illegal or unfair augmentation because this attribute undermines a sport's core tenets.[13] What binds this core together is the belief that unaltered athletic bodies should be the final determiners of winning and losing. These sporting axioms state that even though human and technoscientific judging systems may occasionally fail, the best bodies—and *not* the best technoscientific implement or team—should still determine the outcomes of athletic competitions.

For Pistorius's supporters and critics, whether his prostheses illegally enhanced his natural ability, repaired his disabled body to make him "normal" again, or gave him assistance during competition is central to the concerns regarding his ability to run, and what permitting him to run means for the future of track and field and sport in general. The issues surrounding bodily repair, assistance, and augmentation are at the heart of many legal battles regarding the use of technoscience in sport. In June, 2016, the International Association of Athletics Federations denied German long jumper Markus Rehm—who uses "a long jump–specific prosthesis" in competition—his bid to be the next prosthetic limb–wearing athlete in the Olympics because he was unable to prove that his prosthesis did not provide him with an unfair advantage.[14] The International Association of Athletics Federations has used the language of "improper" technical devices as a way of limiting the use of technoscientific tools for competition. What makes the language of augmentation so volatile is that it taps into fears about unfair additions to an athlete's physical body. The concern is that an unfairly augmented athlete will perform above and beyond his or her natural ability. Rehm represents this anxiety because he potentially is a more dangerous athletic threat than Pistorius ever was because he would have been a medal threat at the 2016 Rio Olympic Games. His personal best of 8.4 m would have won the gold medal at the 2012 and 2016 Olympic Games. The balance among repair, assistance, and augmentation, and the competitive impacts of each, often is very difficult for scientists, governing bodies, and, most importantly, the public to discern, let alone fully grasp.[15]

Pistorius's extraordinary sporting records have been viewed, until recently, as awe-inspiring, celebratory achievements. What were seen as heroic athletic accomplishments within the Paralympics stand in marked contrast to the suspicion imposed when Pistorius's triumphs were reimagined within an Olympic context. Pistorius's potential as an Olympic athlete incited accusations about augmentation, performance enhancement, and unfair advantage. The future of sport itself was imagined as being threatened by a post-Pistorius, technoscientific takeover. Indeed, the blurred distinctions between technoscientific therapy and enhancement, humans and machines, and the Olympics and Paralympics fueled such alarmist rhetoric. Pistorius, prior to his arrest and subsequent conviction for murdering Reeva Steenkamp, presents an excellent case to examine the nature of technoscience and the migrating perceptions of sporting ability and disability.[16]

The Making of the Blade Runner

Oscar Pistorius was born on November 22, 1986. He was diagnosed at birth with fibular hemimelia, a condition that produces shortened fibulae. Before he was a year old, his parents made the decision to amputate both of his legs below the knee. He learned how to walk using prostheses and stated that he has no memory of his body other than with prosthetic limbs. Not surprisingly, he had a very physically active youth. Pistorius notes that he competed in rugby, water polo, tennis, and wrestling throughout his early life. A rugby injury in the summer of 2003 led him to track and field. At the University of Pretoria's High Performance Centre, he began track workouts as part of a rehabilitation regimen for the rugby accident. A lifetime of athletic fitness enabled him to easily transition into competitive running in January 2004. He developed quickly into an elite runner, and by the summer of 2004 he earned a spot representing his home country of South Africa at the Athens Paralympics. By all standards, the seventeen-year-old Pistorius had an excellent Paralympics. He won the bronze medal in the 100 m and the gold in the 200 m, setting a new world record with a time of 21.97 seconds.

At the time, Michael Johnson held the able-bodied world record with a time of 19.32 seconds. Though a 1.5-second difference is sizable in elite track and field, it was an amazing, and for many an incomprehensible, accomplishment for an athlete running with prostheses. Sport journalists began actively comparing Pistorius to able-bodied competitors—even remarking that Pistorius would have won the 400 m final at the 1928 Summer Olympics in Amsterdam—and speculating how many years it would take to close the gap between able-bodied and disabled athletes.[17] Pistorius seemed the most likely individual to cross over and compete with, and beat, able-bodied runners. In hindsight, the IAAF undoubtedly took note of his noteworthy Paralympic performances and what an athlete like Pistorius could mean for the future of track and field—and potentially the

Olympic Games. Public questions about the degree to which his prostheses augmented his performances did not start with the IAAF, the International Olympic Committee, or the Athletics South Africa (ASA) but began with his fellow Paralympic competitors in 2004.

The origin of this controversy begins with Pistorius's prostheses. Fitting prostheses is as much an art as it is science,[18] and the fit of prostheses for everyday use is decidedly different than that for elite-level track and field athletes. The specific prostheses have different functions, similar to the distinctions between casual shoes and track spikes. When Pistorius began running, he worked with South African prosthetist Francois Van Der Watt. In 2004, the market for racing prostheses was small and the science and engineering was just starting to come together to develop decidedly fast limbs. Van Der Watt also was relatively inexperienced, having received his American Board for Certification in Orthotics and Pedorthics in 2004. When Van Der Watt sourced carbon fiber racing prostheses from a local engineer in the aeronautics business, the limbs proved to be structurally unreliable. To resolve this problem before the Athens Paralympics, Pistorius and Van Der Watt arranged a meeting with Brian Frasure. Frasure, a former collegiate runner who lost his feet in an unfortunate train-hopping incident in 1992, dominated Paralympic sprinting competitions from 1997 through 2003. He finished his career in 2008 with thirty major victories and eight world records.[19]

For Pistorius and Van Der Watt, it also helped that Frasure was a prosthetist. It was Frasure who chose to fit Pistorius with the cutting-edge Össur Flex-Foot Cheetahs that led to the press christening Pistorius as the "blade runner." The Össur carbon fiber prosthetic limbs—pioneered by inventor Van Phillips—proved to be a monumental leap forward for competitive track and field athletes for their flexibility, lightness, and strength.[20] In fact, when athletes first competed using the new technoscientific devices at the 1998 Paralympic Games, the 100 m world record dropped almost 1.5 seconds to 11.73 seconds.[21] This is all the more astounding considering the progression in the able-bodied 100 m. The IAAF first began keeping official 100 m world records in 1912, which American Donald Lippincott held with a time of 10.6 seconds. In 1998, Canadian Donovan Bailey held the record at 9.84 seconds. Thus the time progression from 1912 to 1998 was .76 seconds, approximately half of what had been slashed from the record at the 1998 Paralympics Games. This transformative increase in performance can only be attributed to the use of competition-altering technoscience. It was clear to all participants and viewers that the new technoscientific device ushered in a new era in the Paralympics and disabled track and field, and alluded to the potential of competing with able-bodied athletes.

Part of the art and science of fitting racing prostheses, or any other prosthesis, is to determine the "natural" length of the limbs being replaced. For Pistorius, they used a combination of "wingspan and femur measurements to come up with a conservative estimate of Pistorius' anatomical height."[22] Though substantive research undergirded the development of estimates based on the relative lengths of existing bones, it still was an estimate. It is impossible to determine what Pistorius's "accurate" height would be since his lower limbs were amputated at such an early age. When Pistorius arrived at the Paralympics, some of his competitors felt that he was "running tall," or embellishing the estimate of his unamputated height by using longer prostheses. The advantage of running tall is that an athlete takes fewer steps than his competitors due to increased height and stride length. Thus if Pistorius possessed a similar turnover to his competitors, he would cover more ground in the same time and thus get to the finish line quicker.[23] In this case, the prostheses that enabled him to run tall potentially supplied him with an unfair technoscientific advantage. This is no small issue in track and field. Researchers Ralph Beneke and Matthew J. D. Taylor have shown that stride length is one of the keys to Usain Bolt's sprinting success.[24]

Initially, the International Organizations of Sports for the Disabled devised a formula—length of thigh bone minus 13 centimeters, divided by .4—to determine the maximum legal bilateral amputee leg length. Currently, the International Paralympic Committee (IPC) uses a three-step process. The first step is to "estimate maximum standing height from Ulna length." Step two is to "estimate maximum standing height based on measurement of Demi-span," which is the wingspan from the tips of the middle fingers. The final step combines the values from the first and second steps to determine an athlete's maximum standing height by taking "the mean of the two estimates, maximum standing height estimated from ulna length and maximum standing height estimated from demi-span." This estimated height comes with the following caveat: "The overall standing height of the athlete with their competitive prostheses on must be less than or equal to the mean estimated height plus 2.5%. The maximum standing height will be kept on permanent record in the IPC Athletics Sports Management Data System (SMDS) database."[25] These measurement results can have a profound impact on a competition. The question of Pistorius running tall did not persist, but interestingly Pistorius and the issue of running tall would resurface during the 2012 Paralympics. But in a turn of events, Pistorius leveled the accusations that a fellow competitor ran tall, garnered an unfair advantage, and was not in compliance with the rules.

Pistorius's success at the 2004 Paralympics propelled him into the public spotlight, and the young and charismatic athlete became a beloved hero. His success prompted questions about whether he could, would, or should compete against able-bodied athletes. Pistorius did nothing to dissuade the belief that he was capable of competing. He confirmed his competiveness in 2004 when he won an IAAF-sanctioned open competition in Pretoria, South Africa. He received invitations to compete in IAAF international events in 2005, but decided against it. That would change on March 17, 2007, though, when a focused and fit Pistorius received a silver medal in the Senior South African Championships for able-bodied athletes. He quickly went from a heartwarming human-interest story to a legitimate contender to make the South African Olympic team for the 2008 Beijing Summer Olympics. His term as an Olympic contender was short lived, however. On March 26, 2007, the IAAF Council met in Mombasa, Kenya, to deliberate on proposed changes to the competition's rules. The council reviewed 143 proposals but only chose to make one change, which became known as Rule 144.2(e). The council directed this rule change at competitive "technical aids." Specifically, the rule banned the "use of any technical device that incorporates springs, wheels or any other element that provides the user with an advantage over another athlete not using such a device." The new ruling concluded by fine-tuning its language to prohibit the "use of any appliance that has the effect of increasing the dimension of a piece of equipment beyond the permitted maximum in the Rules or that provides the user with an advantage which he would not have obtained using the equipment specified in the Rules."[26]

This ruling, a mere nine days after Pistorius's success at the Senior South African Championships, reeked of discrimination and seemed to speak specifically to prosthetic limbs. In fact, many commentators, such as

Jeré Longman of the *New York Times*, suspected that the IAAF devised the rule to derail Pistorius's Olympic bid.[27] The IAAF denied such a motive and argued that it intended for the rule to address "the use of spring technology in running shoes."[28] But issues regarding shoe technology had been around for some time and had been addressed in the existing Rule 143.2, which stated that "shoes . . . must not be constructed so as to give an athlete any unfair additional assistance, including by the incorporation of any technology which will give the wearer any unfair advantage."[29]

The language about springs as a form of advantage seemed to point directly to Pistorius and his prostheses. Furthermore, the IAAF Council usually presented rule changes to IAAF Congress delegates for voting a few months after the council meeting, and the new rules did not go into practice until after a confirmatory vote. In 2004, the IAAF had scheduled the congress meeting for August 23 and 24 in Osaka, Japan. Instead of waiting for this August date and a vote by its delegates, however, the IAAF Council invoked Article 6.11c and "implemented [Rule 144.2(e)] with immediate effect."[30] The IAAF Congress needed to ratify Rule 144.2(e) after the fact, but the IAAF made it abundantly clear that certain technoscience, presumably prostheses, would not see the light of day at the Olympics. As a result of this ruling, the IAAF-sanctioned Norwich Union Glasgow Grand Prix rescinded its invitation to Pistorius for its June 3 events.[31]

In defending its ruling as not being an attack on Pistorius, the IAAF changed its tune slightly. On June 15, when asked about Pistorius and the IAAF ruling at an Oslo Golden League press conference, IAAF president Lamine Diack confirmed that Pistorius "would not be excluded unless the IAAF received scientific evidence demonstrating that his prostheses gave him an advantage."[32] With this blessing from the IAAF president, Pistorius was free to again compete in able-bodied track and field events. He received his next able-bodied event invitation to the Golden Gala meet in Rome on June 25. An invitation ten days after the IAAF president publicly allowed him to compete indicated that his presence was in high demand. The fascination with him doing the seemingly impossible and keeping up with able-bodied athletes was a seductive spectacle. Though only running in the 400 m "B" event, he performed admirably and finished second.

It is also important to note that the IAAF hired an Italian sports laboratory to capture the event in order to discern if Pistorius's prostheses gave him an unfair advantage. The lab used high-definition equipment to record Pistorius running at competition speed. Subsequently, the IAAF used this data to defend its position in the Court of Arbitration of Sport case brought by Pistorius, allowing him to run in the Olympics. For the IAAF, the video showed that Pistorius negative split his races. That is, he ran faster split times in the last 200 m of a 400 m race. This fact was not a significant surprise because it was easy to see that Pistorius had great closing speed. However, it was mildly unusual in that most able-bodied 400 m runners tend to positive split their races and run faster in the first 200 m. Pistorius's split times may have been considered odd, but not alarming because the history of track and field is littered with philosophical and scientific debates regarding the advantages and drawbacks of negative, positive, or neutral split times and pacing.[33] In the CAS case, the IAAF claimed its video analysis was to determine if "Pistorius' stride-length, [or] the length of time that his prostheses were in contact with the ground, was significantly different from those of the other runners."[34]

The analysis confirmed that Pistorius's running did not vary appreciably from the other competitors, or, more specifically, that his prostheses did not act as springs that augmented his running ability. The IAAF charged Dr. Elio Locatelli with the responsibility of determining if Pistorius's prostheses violated Rule 144.2(e). It is unclear if the IAAF believed that Pistorius's prostheses gave him a technoscientific advantage, but all the effort it put into proving that he violated Rule 144.2(e) implies that it viewed Pistorius with great suspicion and concern. Furthermore, the IAAF undoubtedly instituted Rule 144.2(e) to stop Pistorius from being the athlete that would open the Olympics to disabled athletes, thus undermining the unspoken Olympic ideal of "able-bodied physical perfection and performance."

To resolve the issue of prosthetic advantage, Locatelli employed the services of Professor Gert-Peter Brüggemann of the Institute of Biomechanics and Orthopaedics at the German Sport University in Cologne. Pistorius consented to submit himself for biomechanical study on July 24, 2007, and all parties agreed that the testing would take place in Cologne on November 12 and 13, 2007. The IAAF instructed Brüggemann "to evaluate Mr. Pistorius' sprinting movement . . . and to study Mr. Pistorius' oxygen intake and blood lactate metabolism over a 400-metre race simulation."[35] Prior to the testing, Brüggemann suggested revisions to the testing protocol that would replace treadmill testing with bicycle testing. He encouraged this modification after being notified that running on a treadmill could be dangerous for Pistorius. This was seemingly a strange adjustment since the test was all about his running performance.

Pistorius ran a few other able-bodied events during the summer, but his Olympic hopes hinged on the outcome of Brüggemann's tests. When it came time for testing, Pistorius traveled to Cologne with his agent, Peet van Zyl, and Knut Lechler, a technical prosthetist from Össur. The first day centered on comparing Pistorius's running performance against five control athletes with 400 m times comparable to Pistorius's. Brüggemann instructed Pistorius and the five control athletes to run a below-maximal-effort 400 m outdoors followed by a set of maximal and submaximal 100 m runs on the institute's indoor track. The research team recorded VO_2—the volume of oxygen a body uses to produce energy at a cellular level—and blood lactate numbers for all of the athletes. On the second day, researchers recorded the athletes' anthropometric measurements and calculated their metabolic capacity. The final report was issued on December 17, 2007. The IAAF's interpretation of the study, which Brüggemann and his research team subsequently published in the journal *Sport Technology*, would not produce a favorable outcome for Pistorius.[36]

The report drew a set of conclusions indicating that Pistorius's prostheses provided him with a technoscientific advantage and violated Rule 144.2(e). First, the report asserted that "the metabolic tests indicated a lower aerobic capacity of the amputee than of the controls. In the 400m race the handicapped athlete's VO_2 uptake was 25% lower than the oxygen consumption of the sound control." This was a powerful conclusion. It indicated that Pistorius processed one-quarter less oxygen than comparative athletes and suggested that he needed some additional assistance—implying his prostheses—to keep up because his body was less efficient at processing oxygen. The result also could have been interpreted as Pistorius's body being fitter or more efficient at processing oxygen, but the size of the difference alluded to his prostheses augmenting his cellular limitations. It is also necessary to note that pejorative normative assumptions about the comparative performance characteristics between able-bodied and less-able-bodied athletes drove this analysis.

Next, the report noted that Pistorius's prostheses outperformed the human ankle. Specifically, in comparing "the joint kinetics of the ankle joints of the sound legs and the 'artificial ankle joint' of the prostheses . . . , energy return was clearly higher in the prostheses than in the human ankle joints." It also dismissed potential able-body advantages by determining that during running, knee and hip joints "did not demonstrate any advantages for the natural legs in relation with artificial limbs." The report emphatically concluded that prostheses provided an undisputed advantage. "In total, the double transtibial amputee received significant biomechanical advantages by the prostheses in comparison to sprinting with natural human legs. The hypothesis that the prostheses lead to biomechanical disadvantages was rejected. Finally it was shown that fast running with the dedicated Cheetah prostheses is a different kind of locomotion than sprinting with natural human legs. The 'bouncing' locomotion is related to lower metabolic cost."[37]

After the report was released, it took very little time for the IAAF Council to reach the inevitable decision on Pistorius and his prosthetic limbs. In its press release of January 14, 2008, the IAAF summarized the most damning components of the Cologne report and "decided that the prosthetic blades known as 'cheetahs' should be considered as technical aids in clear contravention of IAAF Rule 144.2(e). As a result, Oscar Pistorius is not eligible to compete in competitions organized under IAAF Rules."[38]

Nearly as quickly as Pistorius rose to public attention, the IAAF attempted to banish him from able-bodied track and field, and, most importantly, the 2008 Olympic Games. The IAAF reestablished the balance between track and field and technoscience by placing the ruling into effect immediately. But this equilibrium was short-lived—Pistorius was not about to let his Olympic dream fade into the distance with what he saw as a contrived effort to limit him, his body, and his prosthetic limbs to unsanctioned running exhibitions and the Paralympics.

The IAAF, the CAS, and the Future of Prostheses

The IAAF ruling implied that there was something inherently wrong with Pistorius's body—that it was not normal, not sound. Most importantly, he—as a new cyborg runner due to the combined function of his body and prostheses—was someone to be stopped. Pistorius did not go away quietly. He, with the assistance of leading sport attorney Jeffrey L. Kessler, appealed the decision to the CAS on February 13, 2008. Pistorius requested that the CAS overturn the IAAF's decision and reinstate his competition privileges. The case moved quickly because of the immediacy of the situation. Pistorius submitted his brief on March 25, and the IAAF submitted its reply brief on April 25. The CAS heard the case on April 29 and 30.

Pistorius raised four major concerns with the IAAF's decision. His legal team questioned the following, focusing on due process, jurisdiction, discrimination, and Rule 144.2(e):

1. Did the IAAF Council exceed its jurisdiction in taking the IAAF decision?
2. Was the process leading to the IAAF decision procedurally unsound?
3. Was the IAAF decision unlawfully discriminatory?
4. Was the IAAF decision wrong in determining that Mr. Pistorius's use of the Cheetah Flex-Foot device contravenes Rule 144.2(e)?

Pistorius's team started by commissioning its own biomechanical and physiological study to verify the conclusions of the IAAF-sponsored investigation. The researchers included Peter Weyand, a physiologist whose research involved force during locomotion (and whose treadmills could calculate sprinting force); Matthew Bundle; Craig McGowan; Alena Grabowski; Mary Beth Brown; Rodger Kram, an expert in biomechanics; and Hugh Herr, a double amputee and a well-known biophysicist.[39] The team aimed to produce a more scientifically rigorous analysis than what they saw in the Cologne report. Their research would eventually be published in the *Journal of Applied Physiology*.[40] This report, known as the Houston report, drew opposite conclusions from those of Brüggemann and the Cologne report.

In presenting its case to the CAS, Pistorius's legal team focused on two key issues: (1) whether Pistorius's prosthetic legs provided the biomechanical advantage of running more efficiently than an able-bodied athlete and (2) whether Pistorius experienced a metabolic advantage because he did not have to provide energy to limbs that the prostheses replaced. For running efficiency, the Houston report concluded that a runner's stride is highly individualized. Specifically, Pistorius's running style had very little bounce. His unbouncy stride could be seen as giving him an advantage because he would lose less energy per step while maintaining a superior mechanical advantage through his prostheses. In a sense, a prosthesis can be more efficient than a human ankle if this movement is disaggregated and abstracted from running as a complete bodily process.

The Houston report asserted that a complex set of physiological and trained movements contributed to a runner's speed. This set of actions made it exceedingly difficult to ascertain if one type of sprinting was more advantageous than another. The Cologne report emphasized that the springlike action of a prosthesis gave an athlete wearing the device an advantage, whereas the Houston report noted that placing too much emphasis on the prosthesis ignored the complex set of bodily interactions that enabled a body to walk or run. The Houston study also showed that Pistorius did not receive a metabolic boost because he was missing lower limbs. Furthermore, if the prostheses gave him an unfair technoscientific advantage during running, this advantage would most certainly have appeared in metabolic studies showing that he used less oxygen than competitors during sprinting.[41]

Pistorius's legal counsel also highlighted some of the significant problems with the IAAF's ruling. First, they found the test results to be misleading. Specifically, the IAAF requested that Brüggemann report on Pistorius's running at full acceleration. This overlooked what Brüggemann, the Rome video, and everyone who had watched Pistorius run clearly knew: he was a slow starter. He was a slow starter because his prostheses did not provide him with the explosive power needed to start quickly. Undoubtedly, the IAAF knew this as well but chose to overlook this reality, its impact on his sprinting performance, and how it dictated the imbalances in his race splits. Secondly, the IAAF's decision to exclude Dr. Robert Gailey—Össur's and Pistorius's scientific expert—during the entirety of the Cologne testing was highly questionable. Gailey, from the University of Miami Medical School, possessed decades of experience studying amputee track and field athletes.

The omission of Gailey reinforced the perception that the IAAF wanted to manipulate the results to negatively impact Pistorius. Locatelli deliberately undermined Gailey's efforts to participate. Gailey emailed Locatelli on October 29, 2007, with a "number of questions and suggestions directly relating to the testing protocol." Locatelli never transmitted this correspondence to Brüggemann. In the end, the IAAF so effectively excluded Gailey and Össur from the process that Gailey "declined to attend the Cologne tests."[42] More troubling is that Brüggemann never saw or approved the final report submitted to the IAAF used to ban Pistorius. Brüggemann, after reviewing the summary during the CAS case, later commented that the summary "was not wholly accurate."

Finally, the CAS case illustrated that "some IAAF officials had determined that they did not want Mr. Pistorius to be . . . eligible to compete in . . . IAAF-sanctioned events, regardless of the results that properly conducted scientific studies might demonstrate." In voting for or against Pistorius's suspension based on the modified summary of the Cologne report, the IAAF sent the relevant files to council members on Friday, January 11, 2008, and the IAAF expected the members to return their votes by Monday, January 14. Because of this brisk turnaround, fourteen of the twenty-seven council members did not return their votes in time. The IAAF extended the deadline, but the CAS case noted that abstentions counted as a vote for banning Pistorius. Pistorius's legal team powerfully demonstrated how the IAAF manipulated the scientific analysis and its own evaluative system to make sure that Pistorius would not be allowed to compete "without attempting to seek any alternative solutions, modifications, or adjustments that might permit him to participate in . . . events on an equal basis with all able-bodied athletes."[43]

The panel adjudicating the case agreed with a majority of Pistorius's arguments and ruled in his favor on May 16, 2008. It was a monumental victory for Pistorius, but not necessarily for other runners using prostheses. As important as the decision turned out to be, there were equally important aspects concerning athletics and prostheses that the case did not answer. In particular, the CAS ruled that Pistorius "is currently eligible to compete in IAAF-sanctioned events while wearing Össur *Cheetah Flex-Foot* prosthetic model as used in the Cologne tests and presented as an exhibit in the hearing of this appeal." This was a serious limitation. The CAS ruled on his behalf but demanded that he wear the same model of prosthesis presented in the case. As a result, the CAS panel determined that its ruling did not apply to any other athletes or any other models or brands of prosthetic limbs. The final reservation centered on the development of new testing methods. At the time of Pistorius's testing, complete technoscientific analyses did not exist to fully confirm or deny if his prostheses gave him advantages or disadvantages over feet, ankles, fibulae, tibiae, and the associated tendons, ligaments, muscles, nerves, veins, and arteries. But the CAS decision did not rule out the potential of retesting Pistorius at a later date, when the technoscience became available. It maintained the right to reverse its judgment and reinstate the ban if it ever was proven conclusively that Össur's Cheetah Flex-Foot prostheses significantly augmented his athletic ability.[44]

So Pistorius won this battle, but his journey was far from over, as he now had to refocus his sights on qualifying for the 2008 Summer Olympic Games in Beijing. To make the South African team, he needed to be one of the top four 400 m sprinters selected from his country and he also had to run the Olympic "A" qualifying time of 45.55 seconds, faster than he had ever run before. This was not an insignificant task. Pistorius would have to run a personal best, which in theory he could attain, but his training over the previous few months had not been ideal for such an endeavor. Though he could have been selected for the 4 × 400 m relay team without the Olympics "A" qualifying standard, Pistorius still had to run faster than 46.02 seconds to be considered. He made three attempts—on July 2, 11, and 16. It did not help the image of the IAAF that, between the second and third meets, IAAF general secretary Pierre Weiss commented that the IAAF preferred that the South African Olympic Committee not select Pistorius "for reasons of safety." Nick Davies, the IAAF's spokesman, defended Weiss and the organization by contending that in track and field "the relay is a scrum" and that he was concerned about the potential of Pistorius doing "serious damage" to his fellow competitors. He reiterated that the IAAF would abide by the CAS ruling but implied that if Pistorius were to make the qualifying time, then his prostheses must be providing the extra power.[45]

In the end, Pistorius ran a personal best of 46.25 seconds on his last attempt, but four other South African sprinters had run faster, of which Athletics South Africa had selected two to compete in the 2008 Olympic Games.[46] Determined to keep pushing forward, Pistorius set his sights on making the 2012 South African Olympic team. In the summer of 2011, he ran faster than 46 seconds three times, and, astonishingly, on July 19, at the Atletica Sports Solidarity Meeting in Lignano, Italy, he ran his all-time personal best of 45.07, which was the fifteenth fastest time of the year.[47]

Athletics South Africa selected him for their 2011 World Championship team and, subsequently, for its 2012 Olympic team on July 4, 2012.[48] Though on the surface it appeared that the Pistorius story was one of success, questions regarding whether prostheses unfairly enhance an athlete's performance were far from settled.

In attempting to discern what constituted an unfair advantage, the CAS panel ruled Pistorius eligible because their testing showed that the Cheetahs offered no *net* advantage. The CAS panel took aim at the drafting of the Rule 144.2(e) and called it a "masterpiece of ambiguity." Specifically, it found the term *technical device* particularly problematic. In a sporting world replete with technoscientific devices designed to give athletes every advantage possible, the broad definition of *technical device* possessed very little explanatory power. Furthermore, there seemed to be a large gap between active and passive technological devices, and it was not clear if and how Össur's Cheetah Flex-Foot prostheses fit into either category. The IAAF indirectly argued that they were springs, which violated the specific language of Rule 144.2(e), but the CAS Panel asked, "what constitutes a device that *incorporates springs*?" The panel continued to deconstruct the specific language in the rule by conveying that "almost every non-brittle material object is a 'spring' in the sense of it having elasticity. Certainly the *Cheetah Flex-Foot* is a 'spring', but does it incorporate a 'spring'? A natural human leg is itself a 'spring.'"[49]

The CAS panel exposed the ridiculousness of the rule's broad claims and honed in on the problematic way the IAAF construed advantage. The IAAF viewed advantage as absolute. That is, any technical device that provided *any* athlete with an advantage, no matter how small, over another competitor was in violation of Rule 144.2(e). The panel firmly disagreed with this interpretation and concluded that it was preposterous to surmise that Cheetah Flex-Foot prostheses transgressed Rule 144.2(e) without appropriate and convincing scientific evidence. The CAS panel strongly rejected the IAAF's perspective that any advantage, not a net overall advantage, was grounds for prohibiting Pistorius from competing and found this reasoning to fly "in the face of both legal principle and commonsense." The inherent problem with the IAAF's position was that it conflated all non-prostheses-wearing bodies. It did not acknowledge that all athletes' limbs are different. All legs do not perform similarly. Some have more bounce than others. Some are stronger than others. Thus the IAAF pitted Pistorius and his prosthesis-wearing body against a fictional norm that the IAAF made sure he was seen as undermining. The CAS panel did not subscribe to this narrow interpretation and chastised the IAAF's Locatelli when he articulated that Rule 144.2(e) would not prohibit Pistorius from running in 100 m and 200 m events, races in which he was at a marked disadvantage because of his slow start. Since the IAAF did not ask Brüggemann to evaluate overall net advantage, this critical analytic component was missing from the Cologne report. The CAS panel firmly, if not narrowly, decided on Pistorius's behalf, but the scientists who produced the Houston report were not in complete agreement regarding whether Pistorius gained an advantage from his prostheses, either.[50]

Revealing the Power of Prostheses

Though research studies had been performed on amputee runners prior to the CAS case, Pistorius and his artificial limbs created a vigorous research debate regarding the extent to which his prostheses did or did not unfairly enhance his running ability.[51] The first disagreements about Pistorius came from Peter Weyand and Matthew Bundle, who believed that Pistorius's prostheses gave him an advantage over able-bodied runners. Conversely, Hugh Herr and Rodger Kram found that prostheses inhibit as much, if not more, than they help. The *Journal of Applied Physiology* chronicled a version of the debate in 2010. Weyand and Bundle argued that prostheses allowed for faster limb repositioning. What this means is that Pistorius has a higher turnover rate. In particular, they noted that "the stride frequencies attained by our double amputee sprint subject at his top speed were greater than any previously recorded during human sprint running of which we are aware." In their lab study, Weyand and Bundle estimated that Pistorius's turnover was 15.8 percent faster than the able-bodied athletes tested. They determined the swing of his leg at his top rate of speed to be 17.4 percent quicker than the first and second place finishers in the 100 m final at the 1987 World Track and Field Championships. As a result of their analysis, they determined that prostheses can artificially augment a runner's ability and concluded that "the moment in athletic history when engineered limbs outperform biological limbs has already passed."[52]

Kram led a response from Alena Grabowski, Craig McGowan, Mary Beth Brown, and Hugh Herr, in which they countered that Weyand and Bundle based their claims on very little data. They revealed that "a grand total of $n=7$ metabolic running economy values for amputees using RSP [running specific prostheses] ha[d] been published. Even worse, ground reaction force (GRF) and leg swing time data at sprint speeds exist[ed] for only one amputee, Oscar Pistorius." Kram and his co-authors noted that Pistorius's leg swing was fast and the light weight of carbon fiber prostheses contributed to the speed but that he was unable to create as much power or GRF with each step as an able-bodied athlete. In sum, they concluded that Pistorius had a quick turnover rate but did not create as much propulsive force as an able-bodied athlete.[53]

The biomechanical debates about net advantage are far from over because one athlete "may use his hips more than the next. Another may rely more on his arm thrust. Amputees develop ways to interact with their prostheses that makes sense for them."[54] The debate about the technoscientific influence of prosthetic limbs will certainly continue, but one must acknowledge that there was something special about Pistorius in his prime. Racing prostheses have been around for decades, but no one had come close to running as fast in a 400 m race as Pistorius until very recently.

With the CAS opening the door, Pistorius delivered successful, but non-medal-winning, runs at the 2012 London Summer Olympic Games. He seemed poised to quell all concerns about the technoscientific advantages of prostheses. With the Paralympics following closely on the heels of the London Summer Olympics, and hosted at the same venues, Pistorius appeared to be on the cusp of becoming a larger-than-life crossover athlete, bridging the gap between disabled athletes and able-bodied competitions. But unfortunately for Pistorius, a strange and ironic twist unfurled after his loss to Brazilian Alan Oliveira in the 200 m final of the Paralympic Games.[55] Media outlets across the globe quoted and requoted a sour Pistorius after his silver-medal-winning performance. "Not taking away from Alan's performance, he's a great athlete, but these guys are a lot taller and you can't compete [with the] stride length. You saw how far he came back. We aren't racing a fair race. I gave it my best." Pistorius ungenerously inveighed that Oliveira's closing 100 m were "ridiculous."[56] This less-than-charitable explanation of his loss prompted Ebenezer Samuel of the *New*

York Daily News to contended that "Oscar the Inspiration became Oscar the Grouch."[57]

The comparison to the surly *Sesame Street* character was not flattering. Pistorius's loosely veiled insinuation that Oliveira cheated and gained an advantage from longer prostheses not only damaged his pristine image but also reignited questions about whether his carbon fiber limbs had unfairly augmented his natural ability in the past. The story also gained energy from televisual analyses in which commentators appeared as shocked as Pistorius and continued to remark on how abnormal Oliveira's finishing speed seemed to be. After the race, Pistorius aimed his public attack at Oliveira, but weeks before the Paralympics he had contacted Craig Spence, the IPC's communications director, to relay his concern that an unnamed athlete was unfairly using prostheses that appeared to be too long.

The implication was that one of his competitors was running tall and would have a stride-length advantage. Of course, the irony of this grievance is that Brian Frasure brought the same complaint when Pistorius began running a bit too fast for Frasure's comfort. Spence publicly responded to Pistorius's post-race assertion, stating: "We measured all eight athletes in the call room prior to the race. All eight were legitimate to race. The measurement is based to proportionality of the body. Clearly, we don't want athletes running on stilts."[58] Though they were all legal, Pistorius was not totally off base with his assertions. Oliveira did change to longer blades three weeks before the Paralympics, yet his new prostheses only increased his height to 181 cm. By the IPC's measurements, he was allowed to run as tall as 184.5 cm.[59]

The most astute observer of these events was Ross Tucker. In posting on his blog, *The Science of Sport*, he examined Pistorius's allegation that Oliveira won because he had a longer stride. First, Tucker performed the scientifically straightforward observation of counting steps. In this regard, he found that Pistorius was, in fact, wrong. Oliveira took ninety-eight steps to Pistorius's ninety-two, or roughly 2.0 m to 2.2 m per stride, respectively. Tucker clearly pointed out that stride length was not the only determining factor in sprinting and that Oliveira's "improved performance may be due to the blade length, it may be due to his technical skill, it may be due to his improved strength, it may be due to some weight loss."[60] In winning the final in 21.45 seconds, Oliveira ran exactly one full second faster than his

previous personal best recorded at the 2011 Brazilian National Championships. Oliveira stated that he tried the longer prostheses in 2011, but that
they did not feel right.

Then, in early 2012, Oliveira tried them again, and three weeks before the Paralympics he decided to make the switch. Oliveira responded
to questions regarding the switch by arguing: "The prostheses don't run
alone. Of course they are good for an improvement but there is not a significant time difference."[61] Pistorius clearly had a vastly different interpretation. Ciro Winckler of the Brazilian National Team contended that
Oliveira's "improvement was also the result of better coaching and facilities in Sao Paulo, together with the fact he had matured physically."[62] It is
difficult to assess how Oliveira had made such substantial gains. The evidence does not necessarily point to the specifics of the prostheses but,
rather, seems to point to the choice of prostheses length.

If Oliveira's new limbs made the difference between a gold medal
and no medal at all, then technoscience does make all of the difference.
But it is important, again, to acknowledge that the art and science that
goes into limb length selection is mostly hidden. Most elite runners used
similar versions of sprinting prostheses to those developed by Össur.
Many athletes also had additional technical assistance from companies
such as Nike (figure 3.2). So, it is not the technoscience in itself that should
be seen as the problem. It is the selection of the prostheses length and the
estimation of height that is at issue. Athletes can select any limb length
they choose so long as it does not surpass the maximum height limit, and
most elite athletes do not compete at maximum height. Tucker rightfully
noted that "athletes settle on a racing height that is lower than the al-
lowable limit for a reason—best performance."[63] Athletes make equipment choices based on engineering data and scientific research but also
based on contractual agreements, personal preferences, and, as Oliveira
noted, "feel."

The Paralympic Games, just like the Olympic Games, is about competing and winning. Each allows athletes to tinker, experiment, and investigate the best equipment choices for a given competition. If Oliveira found
that he was capable of marshaling his body and prostheses to run faster at
a legally taller height, then he was more than able to do so. Maybe Oliveira
was able to perfect his running technique at the new height, or maybe
the rest of the world had surpassed Pistorius mentally, physically, and

Figure 3.2. Nike designed the Spike Pad specifically for Össur's Cheetah line of racing prostheses for Pistorius to use in 2012. Össur; courtesy of NIKE, Inc.

technoscientifically. Pistorius, though he complained, had similar options. He could have chosen to run taller in Paralympic competitions, but due to the CAS ruling he would not have been able to use different limbs in IAAF-sanctioned events. This meant that if he wanted to compete in able-bodied track and field events, he needed to extract every millisecond of speed from his approved prostheses. For better or worse, he was tied to the CAS ruling. In this regard, the issues at hand potentially were more social, cultural, and legal than technoscientific. Nevertheless, Tucker, in his concluding observations, made the prescient pronouncement that in the future we will "watch these races wondering if we're seeing the best runner, the most skillful practitioner, the best engineer or the wealthiest athlete win."[64]

What makes Pistorius's case so compelling for fans, intriguing for scientists, and problematic for sport governing bodies is that he represents our technoscientific hopes and fears all rolled into one individual. In the best cases, technoscience allows humanity to transcend the limitations of human bodies and the natural world to display humanity's creativity, ingenuity, and brilliance. In the worst instances, technoscience is unleashed to undermine the body and the natural world, displaying humanity's shortsightedness, hubris, and arrogance. Legal scholar Shawn Crincoli insightfully indicates that Pistorius falls "in the middle of the cultural uncertainty we share when it comes to science, technology, and difference."[65] The athlete who overcomes all odds is a beloved story, and Pistorius's case takes that narrative to a higher level. He not only overcomes social and cultural prejudice, but the betrayal of his body and nature as well. Yet this narrative is haunted by the question of whether it was humanness or machineness that prevailed.

There are many who have staked out positions on either end of this spectrum, but the reality exists somewhere in the middle. Of course, this can be troubling because, as Crincoli astutely observes, dominant sporting narratives depend on believing that "the beauty of athletic competition stems from its ability to demonstrate the universality of the human condition, [and that] the sporting venue is one in which athletes, without regard to race, class, nationality or religion, have been able to achieve in ways that once seemed impossible."[66] Pistorius clearly taps into this sporting dream, but his prostheses work against the belief in the primacy of the human body because these devices are not of flesh and bone. Though prostheses are a product of human ingenuity, they are still seen as an augmentation of the human body, which inherently raises the question of whether Pistorius is really running. This question is directed at the heart of what is and is not track and field because most events require some form of "running." Legal scholar Patricia Zettler contends that technoscience has been kept at arm's length within sports such as track and field by "mandating historical continuity, promoting 'natural' athleticism, and determining whether a given activity 'counts as running.'"[67] Pistorius's body undermines these agendas.

For historical continuity, Pistorius seems to raise questions about how to compare his results to past performances. There are defined metrics to compare him to other double amputees, but how does his running compare to able-bodied athletes? This is, of course, where issues about augmentation and advantage become historically important and equally contentious. So much of sport is based on comparative analyses of athletic performances, but for Pistorius what is the constant or coefficient by which his performances can be compared to those in the past? Arguably, they could be compared one-to-one, but the visual signification of his prostheses demand that he not be compared equally. If a metric for comparing Pistorius does not exist, his performances are historically immeasurable beyond noting that they are amazing for an amputee. Thus there is conceivably no congruous way to quantify his running, his ability, or his performances. He is the first sprinter to compete at the Olympics using prosthetic limbs. This mere fact makes it hard to compare him to other athletes. A perfectionist narrative supports the belief that athletics is a public display of the capabilities of the human body. But Pistorius's success undermines historical beliefs that athletics is about displaying the triumph of the natural body. There is no place for a person without fibulae and tibiae within this narrative. Pistorius and his performances make a strong case for body perfection, but the public still sees his prostheses as something other than organic bodily material, and it is this distinction that causes concern.

The visually discernable machinelike functionality of prostheses can make observers pause. The athletic augmentations that we do not see—such as reparative surgeries including rods, pins, and screws—enable the public, governing organizations, and athletes themselves to pass as authentically natural. Pistorius is unable to participate in this cognitive sleight of hand. It is the machine aesthetic tied to machine efficiency that makes the viewing public believe his prostheses must do more than repair or assist his incomplete body. Even if the naked eye could not determine if he was using prosthetic limbs, the same question would still persist, is he really running? Though the CAS has judged that he runs, comments by Michael Johnson and others imply that not everyone is in agreement with the ruling of the CAS. If the use of prostheses is understood as repairing his body, then he is conceivably running naturally. But if his prostheses are recognized as assisting his body, then he is running with some help.

As technoscientific studies have shown, if his body is being augmented, then it is unclear if it is Pistorius's body or the machines that are responsible for his elite-level performances. In the context of sport, we are far from embracing the fusion of man and machine to reach a social and cultural location where the cyborg athlete is commonplace. Though Pistorius and others are pushing sporting cultures to welcome this type of body-technoscientific confluence, more often than not fear drives the perception that we as a society are at a critical turning point where technoscience will supersede the natural body and where athletic competition is no longer about the triumph of mind, body, and spirit but about the unwelcomed intrusion of a technoscientific arms race aimed at undermining the purity of sport. As a result, this construction flips Pistorius, and potentially other athletes like him, from being a disabled runner to being a superabled athlete.

For athletes like Pistorius, the tensions between the competing characterizations of disabled and superabled are far from being resolved. Pistorius and other elite disabled athletes are living examples of the existing problems with the term *disabled*. Science and medicine have historically defined and labeled those who are disabled as being abnormal. Of course, the term *normal* is packed with its fair share of social and cultural baggage, but those who seem to fall outside historically defined parameters are labeled as abnormal. In this way, Pistorius's abnormality is similar to other elite-level athletes whose skills and performances are abnormal, if not queer, to others.[68] This is potentially where the language around limiting Pistorius's participation in sport breaks down. In writing about sport and disability, Ivo van Hilvoorde and Laurens Landeweerd noted that "there is no medical categorization of disabilities that fits smoothly and logically into the context of sport. What is considered a disability in 'regular' life may even become an advantage in the context of elite sport."[69] Sport is replete with abnormally, if not freakishly, tall, slight, heavy, muscular, or fast bodies. Pistorius and similarly abled athletes should be welcomed with open arms, but often it is the technoscientific and cyborg nature of their bodies that causes concern. Pistorius's body was not banned, but, rather, the prostheses that were perceived to give him an advantage were banned.

What is troubling about this dual construction of Pistorius being simultaneously disabled and superabled is how he is defined differently than others with abnormally great athletic ability. Amanda Booher indicates that this distinction has profound consequences, stating: "Pistorius' super-ability not only excludes him from competition, but constructs him as a kind of cheater circumventing the 'true spirit of fair play and equality.'"[70] This is the crux of the argument against Pistorius and other less-abled athletes competing in able-bodied sporting arenas. The implication is that Pistorius is cheating by illegally circumventing the rules, norms, and traditions of sport. It is the undercurrent of the perception of cheating that is and will always be highly problematic. Pistorius's abnormality is not celebrated in the same way as Shaquille O'Neal's physical abnormalities. Both Shaq and Pistorius were born with abnormally special bodies, but the technoscience that allows Pistorius to run created a wide gap between the two for governing bodies, fans, and fellow athletes. Pistorius's prostheses do repair his body. They do assist him in running, but this is conceivably no more unfair than any other technoscience used within athletic competition. Thus his "replacement" limbs force publics, governing bodies, other athletes, and companies, such as Össur, that produce these devices to no longer hide behind the veil of natural performance and deny the power, importance, and value of technoscience in sport. Pistorius, and the mass of prosthesis-using athletes coming after him, demand that we embrace this new and emerging convergence of bodies and technoscience within all sporting arenas.

The Pistorius case exemplifies how meanings about sporting embodiment are routinely contested and negotiated. It demonstrates how such challenges reinforce—through the concept of fairness—an already-presumed idea of what the sporting body is and should be. Pistorius's eligibility to compete in able-bodied events relied on the processes of compulsory able-bodiedness, where all parties involved endeavored to locate and classify his body as normal. But we should not lose sight of Pistorius as yet another in a growing number of examples that expose the failure of the modern sporting paradigm. Indeed, as Pistorius and CAS draw attention to the complexity of science, technology, and sport, they also accentuate the instability of all bodies and, by extension, the inadequate conceptualization of the sporting body that constitutes sport's sacred ground. It seems that sport governing bodies have taken on the responsibility of policing this ground because they have a vested interest in protecting the sanctity of human competition. Ironically, sport governing institutions have begun to rely more heavily on the power and authority of science and technology to protect their sports from unwanted and unwarranted technoscientific ingenuity. For technoscience and sport, Pistorius's prostheses reveal the fundamental tensions in the ways in which sporting cultures view the repair, assistance, and augmentation of bodies.

II EVALUATING BODIES

4

"I Know One When I See One"

Sport and Sex Identification in an Age of Gender Mutability

On the evening of August 11, 2012, Caster Semenya made a late race charge on the final straightaway to win the silver medal in the women's 800 m race at the 2012 London Summer Olympics. The twenty-one-year-old athlete had captured her first Olympic medal, but the road to the final would be littered with social roadblocks and cultural potholes. Raised in the northern South African city of Fairlie, in the province of Limpopo, Semenya showed an affinity for athletics from an early age. She excelled at sports, and her strong physique was an asset to these endeavors, until she reached the elite level of track and field. Then, rather quickly, the body that had enabled her to reach the highest level of the sport became a problem—for its seemingly unwomanly appearance and ability.

The rhetoric about "freakish" bodies producing unnatural performances has been a familiar refrain within modern society and contemporary sport.[1] Yet the popular mythology about sport is that competitions should be fair, and in this fairness the best man, woman, or body should win.[2] Over the past few decades, however, it has been harder for publics, competitors, and governing bodies to willingly ignore the profound ways in which athletes, their bodies, and, subsequently, the sports they love have changed with new and emerging technoscience. It is this turn that makes one wonder if recent technoscience, from blood-boosting pharmaceuticals to feather-light shoes, will make the human athlete, as we currently think we know him or her, obsolete. Are we at the precipice of a conceptual shift in sporting competitions whereby athletes' bodies simply mediate a new and potentially more important set of competitions between scientists, engineers, and designers? Possibly, but in the current fanatical sport moment it is unclear if either consuming publics or sport governing bodies will allow sport to openly make this transition. Nevertheless, as emerging technoscience demands a reconceptualization of what sport is and is not, it is running into a host of historically rooted social, cultural, and political affordances.

For Semenya's sport of track and field, the International Association of Athletics Federations regularly grapples with emerging technoscience but has been ineffective in legislating it away from its playing fields. The historical interweaving of the body and technoscience has created a world in which the cyborg nature of sport is becoming increasingly self-evident. Instead of following debates about the ways in which sport governing entities attempt to protect their brands and the illusion that their competitions are contests between bodies of approved athletes, the early running career of Caster Semenya illustrates why a cyborg ideal might positively transform sport and provide an avenue to consider the ways in which cyborg athletes are "about transgressed boundaries, potent fusions, and dangerous possibilities which progressive people might explore as one part of needed political work."[3]

In a similar way, Semenya and her body can be seen as queering the sport of track and field. Semenya and the questions surrounding her gender identity and seemingly non-normative body are ready-made for thoughtful and sophisticated queer analyses, and many scholars have eloquently taken up this task.[4] This chapter does not deny the value and importance of this work; however, in this study on technoscience and sport, the cyborg metaphor and its conceptualization of the body within multifaceted technoscientific infrastructures is the analytic pathway this chapter follows. From a critical technoscientific perspective, the questioning and subsequent testing of Semenya's body demands that we not only examine the social and cultural formations that expect a clean and tidy sex or gender outcome but confirm that the technoscientific testing itself is a dubious method for producing a verifiable truth. If we question the truth-making outputs of gender or sex verification testing, it is possible to positively deploy this interpretive flexibility to rethink and reconfigure future sporting competitions. More specifically, the key question to ask is, what is at stake for a sporting culture if it abandons long-held constructions of the natural athletic body based on outdated sex differences and replaces and revises it with the contemporarily more relevant cyborg athlete?

Getting a sporting culture such as track and field to make this transition is far from easy. The social and cultural embeddedness of the naturalized and gendered body constitutes a series of historical narratives regarding which athletic bodies a governing institution will license to compete in a given sporting arena. Donna Haraway notes that "the trouble with cyborgs, of course, is that they are the illegitimate offspring of militarism and patriarchal capitalism, not to mention state socialism."[5] The production of athletic bodies, from the post–World War II era onward, is intimately connected to the military-industrial complexes of which Haraway writes. Steroids, amphetamines, and a host of other substances were created to produce "new" cybernetic soldiers, but this research also had a profound impact on athletics.[6] As a result, improved training techniques decreased the performance differences between both male and female athletes.[7] The differences in athletic ability between male and female athletes at a young age are shown to be miniscule.[8] However, most professional and Olympic sporting competitions are divided by sex. Historically, this distinction, and the normative reason for gender-based competitions, was a collectively agreed-upon sporting narrative.

Until recently, the broader public understood gender and sex within the context of sport.[9] It was simple. Sex and gender were the same. There were two genders and sexes, male and female, and they were, for the most part, used interchangeably. Men were genetically and physiologically men, and the same was true for women. Moreover, if there was a question about this distinction, one could use a technoscientific test to resolve the issue. Beginning in the 1970s, widely reported news stories such as that of the tennis career of Renée Richards after sex reassignment surgery began to destabilize sex and gender verification testing within sport.[10] Though the power and authority of sex and gender verification testing has diminished significantly, it is far from denuded.

Testing is a familiar and comfortable form of societal evaluation. Educational systems around the globe use various forms of tests to determine if students have studied enough to progress to the next level of education or adequately complete a set of requirements to prove worthy of a degree. Testing is inherently competitive. In most educational systems, tests also delineate between those who pass and those who fail. Outside of an educational setting, testing is strongly connected to feats of strength. Narratives of Greek gods, Norse Vikings, Japanese samurai, Zulu warriors, and Māori fighters highlight the historical power of physical combat as a test of strength, power, authority, manhood, and, ironically, civility. These tests of strength play out on an international level and play into a host of deleterious assumptions about race, gender, nation, and culture.[11] The fighter Jack Johnson's life exemplifies these multiple tensions.[12]

Much of sport is driven by physically and mentally testing one's body. Cricket and rugby have endurance-based test matches designed to test an athlete's ability to perform at the highest level for an extended period of time. These, like most other competitions, are tests of ability, strength, and fitness. By the early part of the twentieth century, *testing* began to be used as a term to determine who was and was not fit to compete. Being "fit" to compete has meant a host of things. Then, it often meant that women could not compete against men or black men could not compete against white men. Ultimately, it reified into a means of institutionalized sport segregation with unarticulated but actionable social and cultural ideas about masculinity, femininity, fairness, equality, race, sex, nation, and entitlement. It also began as a way to evaluate if athletes competed on a level playing field. Testing became a technoscientifically validated way to determine if the bodies of competitors were "pure." What purity meant depended on whether the tests were aimed at rooting out performance-enhancing substances or rogue chromosomes that would unfairly give one competitor an advantage over another. From the early 1970s onward, each Olympic Games has promoted a claim of better and more rigorous testing to assuage the public fears that valorized athletes are not "cheating." As a result of this vision, the International Olympic Committee organized a team of more than 150 scientists to examine more than six thousand samples during the 2012 Summer Olympic Games.[13] This effort seems to have taken a step backward with the 2016 Rio Olympics, in which the WADA suspended the laboratory scheduled to manage drug testing for the games.[14] Some consider the 2016 anti-doping measures to be the worst seen in contemporary Olympic history.[15]

The troubled realities of testing for sex and gender, in conjunction with the late-twentieth-century conceptual and theoretical developments in gender and sexuality studies, present an excellent case for reexamining sex and gender as defining attributes of sport. In this regard, what would happen to sport and what would athletic competition look like if sporting cultures abandoned sex and gender binaries and embraced the continuums of gender and sex? This transformation could potentially result in more interesting, relevant, and subjectively better athletic competitions.[16] While Caster Semenya's experiences over the past decade can be instructive in examining the physical and mental trauma that gender and sex verification can cause, they can also be an impetus for the next evolution of gender and sex within sport.

Questioning Caster Semenya

Caster Semenya's body, sex, gender, race, and identity have all been questioned at various different times by a host of public and private authorities, institutions, and individuals. In July 2008, the seemingly unknown South African runner competed in and won the women's 800 m event at the Commonwealth Youth Games. The *Telegraph* indicated that it was at this race that rumblings began about her deep voice and masculine features.[17] These questioning reverberations turned into actions when Semenya simultaneously won the African senior and junior championships in July 2009 with record-breaking performances. The IAAF argued that "it was obliged to investigate after Semenya made improvements of twenty-five seconds at 1500m and eight seconds at 800m—the sort of dramatic breakthroughs that usually arouse suspicion of drug use."[18] Although they tested Semenya's body for performance-enhancing substances, the fact that the IAAF also requested that Athletics South Africa perform gender verification testing before the world championships in August changed the public's perception of Semenya.

Most of this information would have remained private, if not for a fax related to the testing inadvertently being sent to the wrong individual. The Australian *Daily Telegraph* led the reporting of the story with the bold claim that Semenya had testes and no ovaries.[19] Unfortunately, the news broke a few hours before Semenya was scheduled to run in the 800 m final at the world championships. Not initially ruffled by the flurry of reporting activity, she won the race and became the world champion with a time of 1:55.45, which was the fastest 800 m women's time of the year (figure 4.1).

Figure 4.1. Caster Semenya at the 2009 IAAF World Championships, Berlin. Semenya's muscular physique and victory in the 800 m final caused many to question the athlete's body and the way our society differentiates sporting competitions based on sex and gender. PA Images / Alamy Stock Photo

This race would be her last competition of 2009. She did not see the track competitively again until July 18, 2010. During this period, the ASA maintained that Semenya was not banned from competition but had to wait for test results from the IAAF.

After Semenya's gender verification testing was leaked to the public, the IAAF and the ASA both attempted to sidestep the maelstrom of questioning and criticism regarding who requested the testing, which gov-

erning organization initiated and performed the testing, who knew the results first and when, and what Semenya did or did not know about the testing and the results of the testing. The first public comments from the ASA
came from its president, Leonard Chuene, who denied that any testing had
taken place before the Berlin World Championships or that the ASA had
approved and scheduled any testing. Leaked reports contradicted his
statements. If the rumors were true, why did the ASA allow Semenya to
compete and win a world championship medal?

In a September 11, 2009, interview with the *Sydney Morning Herald*, IAAF spokesman Nick Davies confirmed that gender verification
tests had been performed but only "to assess the possibility of a potential medical condition which would give Semenya an unfair advantage
over her competitors."[20] In fact, the IAAF had contacted Dr. Harold Adams, the team doctor for the South African athletic team for the 2009
IAAF World Athletics Championships, prior to the world championships
regarding its concerns about Semenya's ability to fairly compete as a
woman. Over the course of the next few weeks, bits of information surfaced that resolved many of the contradictions in the ASA's story among
the key players involved in the testing from the ASA: Leonard Chuene,
Harold Adams, Molatelo Malehopo (ASA general manager), Kakata Maponyane (ASA vice president), Phiwe Mlangeni-Tsholetsane (events manager of team South Africa), and Laraine Lane (ASA board member and
psychologist).

On August 5, 2009, Dr. Adams sent an email to Malehopo and
Chuene containing the following statements: "After thinking about the
current confidential matter I would suggest that we make the following
decisions. 1. We get a gynae opinion and take it to Berlin. 2. We do nothing
and I will handle these issues if they come up in Berlin. Please think and
get back to me ASAP."[21] The confidential matter certainly was Semenya,
and they were plotting to figure out how to manage her situation, as the
IAAF was interested in testing her as well. Malehopo quickly responded, "I
will suggest that you go ahead with the necessary tests that the IAAF
might need."[22] According to these documents, Dr. Oscar Shimange, an ob-
stetrics and gynecology specialist at the Mediclinic Medforum in Pretoria,
performed the tests shortly after the email exchange between Adams,
Malehopo, and Chuene. For most of the period during the questioning of
Semenya, ASA board member and psychologist Laraine Lane remained

silent. The most Lane said about the situation was: "I cannot discuss issues regarding my clients. I can't deny or confirm anything; it would be a conflict of interest if I did that. I would like to help you, but I can't."[23]

Lane eventually came forth in 2013 and confirmed that the ASA did indeed perform gender verification testing on Semenya in early August, but her version differs in one specific way from prior public narratives regarding Semenya. The ASA has always contended that Semenya had been informed about the type and nature of the testing performed on her body; however, Lane noted that Semenya was not properly informed. Lane recounts that Adams requested that she "provide . . . Semenya with supportive counselling."[24] Adams called on Lane because she was a psychologist, but more specifically because, as Lane notes, "both he and Caster were leaving for Berlin the following day, [and] there was no time to arrange for anyone else to see her."[25] Yet Lane attested that Adams had not secured the necessary informed consent from the eighteen-year-old Semenya. Lane conveyed that the all-important "supportive counselling would have been appropriate subsequent to the athlete being informed of the procedures involved in the medical investigation, not prior to her being advised of the tests that would be carried out."[26] As a result, Lane supported Semenya's enthusiasm as she departed for Berlin, but she did not provide the level of guidance that was merited under the deeply questionable circumstances.

According to a report by Adams, the Mediclinic Medforum contacted him on August 15 to inform him that they had preliminary results from Semenya's tests. Adams also noted that Semenya supplied the necessary "written approval" for him to receive the results by phone in Berlin.[27] Adams recollected that the data received over the phone was "not good" for Semenya, but that the official report would arrive in a few days for him to review. With the crucial news verbally transmitted, Adams met with Chuene that same day. Adams recounted that he made a recommendation to the ASA for the removal of Semenya from competition, and Chuene agreed. With what he now knew, Adams did not want Semenya to be tested at the world championships and believed that "the withdrawal would give [them] an opportunity to be able to take this process forward back in South Africa, together with all parties concerned (Ms. Semenya, her family, ASA, the IAAF, Government, and any other approved interested parties)."[28] His rationale for this recommendation rested on his belief that "being tested at the World Championships would not give her enough time to consult extensively and perhaps arrive at a decision to refuse the testing if she felt it would infringe on her privacy and personal rights."[29] Apparently, in an effort to withdraw without much fanfare, Adams even suggested that they bandage a part of Semenya's body to feign an injury. Evidently, this suggestion was presented to Semenya, at which point she broke down.

On August 16, Chuene altered his stance on Semenya. He no longer felt it wise to remove her from the competition. Arguably, there was a significant amount of pressure on him to keep one of South Africa's medal hopefuls in the competition. Chuene expressed these new thoughts to Adams the very same day. Adams recounted that Chuene "changed his mind after consulting with high-powered politicians back home in South Africa, as well as . . . Maponyane. Mr. Chuene said if we withdrew Ms. Semenya what explanation would we give to the politicians back home." Adams believed that they could have withdrawn her quietly because she was relatively unknown, but seeing the eruption in interest after the world championships, Adams's assumption was probably a bit off base. According to Adams, Chuene wanted to take a proactive approach and asked him to arrange a meeting with the IAAF's medical team in order "to politicize the whole thing, and to cause confusion."[30] The goal was to intimidate the medical team. Chuene and Maponyane led the meeting with the IAAF medical team and conveyed that the removal of Semenya was more than a sporting issue; it was a political concern. If the medical team were to test Semenya, the South African government would see this as a political offense. Evidently, the medical team did not succumb to these direct threats. They summarily presented the ASA with two options:

1. That Ms. Semenya could compete at the World Championships, on condition that she accepted that she would be subjected to the IAAF's Gender Verification tests in Berlin, and that if any unfair advantage was detected on the part of Ms. Semenya, she would be stripped of any medal she might have won at the Championships; OR

2. That Ms. Semenya is withdrawn from the World Championships. If this was to be the option exercised, the IAAF was comfortable with ASA handling the matter of the Gender Verification Tests back in South Africa, and a report on the said tests sent to the IAAF.[31]

Chuene selected the primary course, and, not surprisingly, the IAAF tested Semenya the day after their meeting.

In early September, the IAAF confirmed its testing of Semenya and released the following appropriately inconclusive and measured statement: "We can officially confirm that gender verification test results will be examined by a group of medical experts. NO decision on the case will be communicated until the IAAF has had the opportunity to complete this examination."[32] Thus the IAAF did nothing to clear up questions regarding what Semenya did and did not know about the testing, but it appears that she was kept mostly in the dark about the types of tests and the intended uses of the results. This was confirmed on September 19, 2009, when Chuene admitted that he had lied to Semenya about the types of tests performed.[33] The complete details of what transpired beneath the sporting, media, cultural, political, and technoscientific circuses over the following months will probably never be fully disclosed, but when Semenya's attorney, Greg Nott, acknowledged that "direct negotiations with the IAAF representatives, through the mediator, [had] been ongoing for 10 months . . . in Monaco, Istanbul and Paris," it was clear that the resolution of the case was as much about testing as it was about politics.[34]

On July 6, 2010, the IAAF released the briefest of press releases, indicating that "the process initiated in 2009 in the case of Caster Semenya (RSA) has now been completed. The IAAF accepts the conclusion of a panel of medical experts that she can compete with immediate effect."[35] No information was ever released to the public about the testing or how these tests informed decisions regarding Semenya's ability to compete. Of course, the Semenya case raises a host of interesting questions about sex, gender, politics, race, and the body in a sporting context. But can her case push toward revolutionizing how publics, sport governing bodies, and competitors themselves understand the problematic workings of gender and sex within the tradition-laden world of sport?

Redefining Athletic Bodies

The experiences of Caster Semenya illustrate that contemporary sport cultures have yet to develop a respectfully coherent way to understand bodies that do not simply conform to the outmoded binary of male and female. Part of the problem for sport has been the slippage in the use and meaning of "gender verification." The most altruistic reading of sex and gender verification is that it initially began as a method to catch "cheaters." Specifically, the goal was to expose men who chose to compete as women. This narrative gained traction during the 1966 European Track and Field Championship, when six competitors from an Eastern Bloc team withdrew from competition when they learned that they would have to pass a physical inspection by a panel of physicians.[36] Although the IOC stopped compulsory gender verification testing in 1999, and no single man has ever been "caught" impersonating a woman, the IOC and the IAAF replaced the system with case-by-case assessment in 2003 and 2006, respectively.[37] Semenya's questionable body was funneled into this new case-by-case system. But the question still remains, what did this new system aim to ascertain?

Laura Hercher argued that since "Semenya was reared as a girl[,] her genitalia are female [and she] self-identifies as a woman . . . how is it possible that any inquiry could authoritatively declare that she is not a female?"[38] The historically rooted misunderstandings of sex and gender difference by competitors, governing bodies, and publics are at the heart of the gender verification "problem." Scientists, from biologists to genetic counselors, have called into question the idea of genetic verification and have bluntly stated that "human society as a whole is lacking in its ability to deal with disorders of sex development (DSD) at either a social, competitive, legal, or clinical level."[39] The concerns arise when androgens are used to determine a perceived level of maleness or femaleness. Certain conditions such as congenital adrenal hyperplasia (CAH), adrenal tumors, or partial androgen insensitivity syndrome can produce higher androgen levels in women or have other effects that could give a woman a genetic advantage for a specific competition. Studies have shown that women with complete androgen insensitivity syndrome (CAIS) and XY chromosomes may have fewer androgens than the "average" female.[40] Thus claiming that a CAIS woman would have an unfair advantage because of a Y chromosome is highly problematic.

Sport governing bodies have slowly accepted this reality but have now, for the most part, discontinued solely using the Y chromosome as an indicator of sex. Nevertheless, athletics still struggles with how to devise a system that genetically or biologically determines what an appropriate natural advantage is. For track and field athletes, the IAAF has agreed that women with CAH, adrenal tumors, or CAIS can compete as women. This is a fine starting point, but what is desperately missing is an understanding of "which conditions disqualify an athlete from playing as a woman."[41] Semenya's experience could have been an important scientific, social, and cultural exemplar that could have provided important guidance for this issue. However, since no information has been released about Semenya's case other than that she was withdrawn from competition and eventually reinstated, her tragic situation will not shed any light on how non-normative bodies will be treated in the future by sporting competitions defined by a male-female binary.

This fact has also left Semenya and her body in a strange space filled with assumptions and speculations that she can only partially resolve on the track. She did recover enough to win silver medals at the 2011 Daegu World Championships and the 2012 London Olympics, but as exercise physiologist Ross Tucker indicates, Semenya will always be under scrutiny. He explains, "Either she would win convincingly, and the world's athletics followers would say 'She has an unfair advantage, they obviously didn't change anything, and now thanks to her lawyers, no other women can even compete.' Or, if she didn't win her races, the world would say 'This proves that she must have had surgery or treatment.' A catch-22 for Semenya."[42]

Questions about where Semenya's body sits on a continuum of maleness to femaleness will always overshadow her competitive successes or failures. There is much room to speculate on the multiple chemical, pharmaceutical, surgical, or psychological options deployed to push, prod, pull, or direct her body off some indeterminate perch on a gender spectrum to comply with the IAAF's evolving definition of femaleness. Drawing from Donna Haraway and other feminist scholars, Semenya's case also highlights how the female athletic body is a construction with publics, governing bodies, competitors, physicians, and scientists invested in maintaining a carefully calibrated equilibrium of female sex identity. The Semenya case produced a new level of murkiness in the collective understanding of the female athletic body because no one released a statement resolving the Semenya conundrum to bring her body back to a familiar prescribed sex orientation. In a sense, sport governing organizations, competitors, and both casual and enthusiast publics have not been allowed to breathe a collective sigh of relief by consuming a tidy resolution to the

Semenya case and confirm that, sadly, in an athletic arena sex has once again triumphed gender.

Caster Semenya's body upset the balance in many ways. It is striking that she was seen as too masculine in a space where masculine athletic prowess is not only accepted but also championed. Her body had apparently crossed the line from being a svelte and toned biological machine to an overly masculine and questionable, freakishly alienating device. And while many competitors mumbled about Semenya's body, only Italian middle-distance runner Elisa Cusma Piccione would publicly state, "For me she is not a woman."[43] In that moment of female hypermasculinity, her fellow competitors, the IAAF, and the sporting press disaggregated Semenya from her body and constructed it as an indeterminate and inadmissible biological device requiring further examination and review. In this evolving moment of what is and is not too masculine within gender-differentiated sporting competitions, the sport governing body chose an equally troubling adjudicator of this bodily conundrum: technoscience.

Historically, technoscientific testing has been a useful tool enabling sport governing bodies to extract themselves from the social and cultural issues precipitated by delineating what is and is not an admissible athletic body or performance. The entire business of drug testing has enabled sport governing bodies to allow science to determine if an athlete is competing within the defined rules of a sporting competition without fully addressing the varying levels of naturally occurring chemicals or conditions that produce competitive advantages for some athletes in certain sports. The Semenya case initially appeared as if it would be a simple probe into a woman's body to determine if it was real, pure, or authentic enough to participate in sport as a woman.

Unfortunately, in the Semenya case the ways in which the IAAF used the historical power of technoscientific testing to sidestep the social and cultural mechanisms that form gender identity are not questioned. It would appear that most sport governing bodies are not particularly interested in going to the "softer" intellectual domains containing rafts of critical gender studies for explanatory evidence. Instead, the IAAF depended on the truth-manufacturing machinery of technoscientific testing to output a sex confirmation on which the appropriate choices would be made therewith. But from what has transpired over the past few years, Semenya's body presented a snag somewhere that demanded that she be pulled from competition for a period of months until her "flawed" sport-appropriate sex could be "reset" or "fixed."

In the end, why was Semenya seen as a problem to be fixed? First, her body, at least for the short period in which she was not allowed to compete, transgressed the boundaries of what Western athletic society deems to be a female body. In recent years, athletes such as Oscar Pistorius, with his J-shaped carbon fiber prosthetics, have pushed sport into an uncomfortable dialogue about how new and emerging technoscience and bodies no longer fit into the athletic performance binary of able body–disabled body.[44] In a similar way, Semenya's body questions the tensions between male and female. We have colluded to maintain the myth that athletic bodies from a physiological, genetic, and technoscientific standpoint are relatively simple and easy to categorize. This is what has been driving the hand-wringing around Semenya, Pistorius, and the next wave of emerging athletes that will not fit valorized athletic narratives.

These historical narratives are also elements of the multiple backgrounds that motivated the IAAF to legislate and rule on Semenya's sex. Her indeterminate gender identity undoubtedly disrupted the orderly nontechnoscientific media fantasy that has long been a cornerstone of sporting competitions. The collective silence around Semenya's body implies that under the current state of the rules, it is unplaceable, which is the dominant message delivered to the masses not privy to the decisions based on the test results. Thus if her body, and certainly future athletes' bodies, cause so much controversy, why not embrace the cyborg nature of the body? This is not a call to open the floodgates where anything is permissible within sport but is a request that sporting culture find a space to discuss, debate, and theorize a future in which sporting competitions can be based on a more complex and comprehensive understanding of the human body and its relationships with humanity's technoscientific output. What would it mean in practice to embrace the gendered and sexually indeterminate athlete? Of course, in this corporate sporting moment it is nearly an impossible prospect, but reading athletic bodies through contemporary frameworks of genetics, sexuality, and gender demands that we ask this question.

A negatively valenced interpretation of this prospect contends that acknowledging the blurred boundaries between male and female athletes would destroy the history and tradition of sport by allowing that something other than anachronistic understandings of genetics, hard work, and perseverance could determine the outcomes of sporting competitions. But what would it mean if we contemplated a more optimistic analysis and insisted on embracing the multiple modes of athletic bodies? This transformation has the potential to resolve deep-seated social, political, and cultural tensions within sport. Specifically, a cyborg understanding of athletes can be leveraged to reconstruct competitions that are no longer based on sex and gender.

Reconfiguring Athletic Competition

At first glance, it would appear that seeing the body as a cyborg would cause concern, but explaining the workings of the body in mechanical terms has been a common trope since Enlightenment. In the first few sentences of the introduction to Thomas Hobbes's *Leviathan*, he writes: "For what is the heart, but a spring; and the nerves, but so many strings; and the joints, but so many wheels, giving motion to the whole body."[45] Hobbes understood and rationalized the body as a machine. This perspective, reflecting the rhetoric of the life sciences, promotes critical commentary within cyborg literature and even resonates with futurists such as Ray Kurzweil, who are preparing for the moment when humans and machines become unified.[46] Framing the body as a machine has profound implications for sport. What does it mean for athletes, fans, and governing organizations to view the organic material of bodies as perfectible pieces of machinery to be inspected? Jan Rintala argues that when bodies are seen as machines, contemporary sport disassociates them from humanity and produces sporting cultures that dehumanize and alienate athletes from their bodies in the quest for increasingly greater displays of physical performance.[47]

More recently, the dynamics of dehumanization and alienation have become less relevant, as science and technology are seen at worst as necessary evils and at best as requisite interventions to keep sport safe, rebuild injured athletes, maintain an upward slope of human performance, and build and sustain public trust. Sporting cultures expect bodies to perform like machines while maintaining their human qualities. This situation gets messy when publics want and even demand great performances no matter the cost but deny the high level of cyborg enhancement necessary to create these performances. As insightful as Rintala's comments are, his writing does not examine the place where Haraway's writing is so relevant: the place of gender.

Sport competitions have been heavily sexed and gendered from the outset. The earliest Olympics did not allow female competitors, and it was not until the 1960s that more than a few competitions existed for women. Institutionalized concerns about sex, gender, and fairness can be traced back to the 1920s and 1930s, with athletes such as England's Mary Edith Louise Weston being questioned for competing with a less-than-feminine appearance.[48] Although Mary would later undergo gender reassignment surgery in 1935 to become Mark, Weston was never tested while competing as a woman.[49]

Avery Brundage, in his role as United States Olympic Committee president, began the push for sex testing after the 1936 Olympics. Gender testing became mandatory for all female athletes in 1966. This testing used a visual assessment to determine if a female athlete was "really" a woman. In order to verify one's gender, organizations such as the IAAF required athletes to parade naked before testing officials, which may or may not have included a gynecologist. If an athlete passed this demeaning "nude parade," she would receive the necessary "Certificate of Femininity" allowing her to compete.[50] The IOC would not replicate this testing disaster for the 1968 Summer Olympic Games. They instead moved to the perceivably more effective method of testing for the presence of a Y chromosome, and eventually to DNA testing. But all testing eventually came under fire, and the IOC and IAAF suspended testing in 1991 and 1999, respectively, though both organizations left an opening for officials and athletes to request that an athlete be tested. The process avoided the denigration of all athletes in exchange for a more private and seemingly secretive method. But as in Caster Semenya's case, this process did not resolve gender issues, which resulted in the IOC taking a hard-line stance in June 2012 to define who is and is not a woman.

On June 22, 2012, the IOC Executive Board clearly outlined its position on sex and gender before the 2012 London Olympics by issuing the "IOC Regulations on Female Hyperandrogenism."[51] The IOC clearly stated that "competitions at the 2012 London Olympic Games . . . are conducted separately for men and women (with the exception of certain events)." The opening statement of this document outlined that the IOC intended to maintain the existing structure of the Olympics as a series of gender-differentiated competitions. The IOC directed this new regulation at those categorized as having the condition of female hyperandrogenism. Though the document stated that "these Regulations are designed to identify circumstances in which a particular athlete will not be eligible (by reason of hormonal characteristic) to participate in the 2012 Competitions in the female category" and further claimed that "nothing in these Regulations is intended to make any determination of sex," this new set of regulations was *all* about sex. It was all about the parameters that the IOC would use to determine if an athlete's body was appropriately, or inappropriately, female. The IOC attempted to skirt this massive social and cultural issue by implicitly making a distinction between legal sex and biological sex. Though the larger world may agree on varying continuums of sex and gender, for the IOC and the Olympics the binary of male and female is the only option. An athlete may live her life as a woman, but the IOC will now use a new set of metrics to trump the legal, social, and cultural parameters by which femaleness is continually redefined.

Equally troubling is the way that the IOC deputized National Olympic Committees and fellow competitors to police the bodies of female athletes. The regulations demand that "each [National Olympic Committee] *shall*, as appropriate, prior to the registration of its national athletes, *actively* investigate any perceived deviation in sex characteristics and keep complete documentation of the findings, to the extent permitted by the applicable law of legal residence of the concerned athlete" (emphasis added).[52] What this seems to do is pass the responsibility, and subsequently the legal liability, onto the National Olympic Committees. Thus it is a National Olympic Committee's job to observe, monitor, and investigate "questionable" female bodies and ultimately determine if these athletes can be categorized as "real" women in the eyes of the IOC Executive Board.

The final authority to determine if an athlete's androgen levels are too high to compete as a female will reside with an expert panel of at least "one gynaecologist, one geneticist, and one endocrinologist," appointed by the chairman of the IOC Medical Commission. The regulations also allow for "an athlete who is concerned about personal symptoms of hyperandrogenism" to request another athlete be investigated. There is a clause that states that if this request is made in bad faith, the IOC Executive Board "*may* impose sanctions on the requesting person" (emphasis added).[53] In a sense, the IOC has legislated a state of surveillance within women's sport.

Critics quickly questioned what these regulations would mean for female athletes. Many viewed these regulations as a direct response to the Caster Semenya case.[54] The inherent problem is that the IOC has chosen to focus on testosterone as the key marker of maleness. Thus if a female athlete's "testosterone levels fall within the normal range of a man," she can be banned. Yet the "IOC does not reveal what a man's normal levels might be."[55] The expert panel will determine the androgen level that "confers a competitive advantage."[56] The IOC's only role is to develop the regulations and administer sanctions. The process is designed to appear to rely on irrefutable scientific evidence, but in the end an expert panel appointed by a governing body makes the final decision. Each expert panelist relies on his or her expertise, skill, and intuition to make a final ruling.

Eric Vilain, a medical geneticist who participated in the discussion to develop the new regulations, acknowledged that they are imperfect but contends that "you have to draw the line in the sand somewhere."[57] He also states that there is a large gap between the upper level of testosterone in a woman's body and the lower level in a man's. The only time these levels will overlap is when an individual is intersexed. If this is the case, these new regulations are inherently directed at eliminating or significantly reducing the number of intersexed athletes in female-only competitions, because only intersexed individuals will conceivably have testosterone levels in the "male" range. This direct attack on intersexed competitors is highly problematic because, "scientifically, there is no clear or objective way to draw a bright line between male and female."[58]

For the IOC, distinguishing between male and female noninter-sexed athletes is seemingly easy. It is all about testosterone. But testosterone is not the only marker of athletic ability and performance. Individual athletic bodies are complex and often convoluted machines that can betray their highly trained owners in an instant. To place the weight of defining the distinction between male and female, even in an athletic context, on testosterone is highly reductionist. The IOC is ignoring its own funded study that showed that the levels of testosterone in male and female elite athletes has overlapped.[59] The only reason to focus on testosterone is to produce a public-friendly way to distinguish between men and women. Rebecca Jordan-Young and Katrina Karkazis are correct in pointing out that "what is really driving these policies is suspicion of women perceived as gender 'deviant.'"[60] The IOC has made it abundantly clear that it governs sport competitions between men and women. Bodies that do not conform to the way they define this binary are unwelcome. Jordan-Young and Karkazis call it "a gender witchhunt." Sadly, it has gone beyond a witch hunt.[61] Witch hunts attempted to discipline women and curtail deviant behavior, but the IOC is attempting to define what a woman is and ban non-normative bodies. The IOC is taking one more step beyond the social and cultural activity of witch hunting and is attempting to cull any bodies that do not conform to their definition of *woman* and *female*.

Unfortunately, the IOC has chosen to be regressive rather than progressive. Instead of applying social and cultural analysis to their formulation, they have chosen to focus solely on a narrow set of parameters resulting from technoscientific tests. Instead of figuring out a way to incorporate intersexed individuals into Olympic competitions, the IOC has turned its back on the individuals it has labeled as unworthy of contributing to the Olympic movement. During his twenty-year term (1952–1972) as IOC president, Avery Brundage was admired, as well as denounced, for his staunch support of amateurism. He was equally fervent about maintaining a sex-differentiated Olympics. It is ironic that the IOC would enshrine Brundage's quest to stop "gender deviants" from scooping up medals by immorally passing as women. The IOC has made it abundantly clear that only approved male and female bodies can compete in the Olympics. The IOC seems to want to quantify gender for its own purposes. This may appear plausible because the current rise of sport analytics has made it possible to conceivably quantify everything in sport.[62] Science has been so good at quantifying, measuring, and classifying the world, so why would it not work for gender? It is disappointing that the IOC is not working harder to learn and use Caster Semenya's case to initiate new discussions about sex and gender in sport. From the outside, it appears that the IOC took the easy way out and punted on gender.

Feminist movements of the twentieth century, and legislation such as Title IX in the United States, led the way for more sporting opportunities for women. Although more opportunities presented themselves, these competitions remained marginalized, primarily because of the perception that the performances by women were of lower "quality." Scientific studies have shown that the physiological differences between elite female and male athletes are quite small when compared with those in the general population.[63] The largest differences are in strength while the smallest are in endurance, but women continue to close the gap. For example, the 2012 NCAA Women's Basketball Championship displayed an increasingly higher level of play. Yet the defining moment of this tournament took place on March 24, when 6'8" Baylor University junior Brittney Griner threw down a two-handed dunk against the Georgia Institute of Technology.[64] This was the second time, since Candace Parker dunked for the University of Tennessee on March 19, 2006, that a woman has dunked in this college basketball tournament.[65] What made this dunk special and different from Parker's, though, was the way she dunked. Parker's one-handed dunk barely cleared the rim, whereas Griner's two-handed dunk was done with power and authority.

During and after the game, social networks were flooded with dialogue and debate about how Griner had just "dunked like a man." As images and videos of the dunk spread worldwide, questions about her sex and gender began to recirculate around the Internet because she was doing things that female basketball players "should not be able to do."[66] Even opposing coaches were using masculinist language to describe Griner. The University of Notre Dame's coach Muffet McGraw commented, after their loss to Griner and her Baylor team in the 2012 tournament: "I think she's like a guy playing with women."[67] What McGraw and others have refused to acknowledge is that Brittney Griner was not like any other woman playing college basketball. Though Griner did eventually come out as gay, which some have used inappropriately to substantiate questions regarding her perceived masculinity, she was not on the spectrum of athletic ability.[68] She was the *end* of the spectrum. The sporting public's problem with Griner was that they were not used to seeing a female athlete play the way she does in a basketball context.

Griner's body is physically comparable to, if not larger than, those of many men playing professional basketball. In a sport where wingspan and hand size are key attributes to playing center, Griner is second to none. Her wingspan is seven-foot-three-and-a-half-inches, which is a half inch longer than former seven-foot-tall NBA center Andrew Bynum. Her hand is a quarter inch larger than LeBron James's hand.[69] Thus, size-wise, she is comparable to male NBA players. But beyond size comparisons, what Brittney Griner represents is the rapid development of sport infrastructure for women and girls in the past twenty years. In more ways than one, this transition has eclipsed Title IX, which allowed women to play, and has spawned events such as the 2015 FIFA Women's World Cup, where women as elite-level athletes displayed the highest level of athletic skill and marketability. Women with elite-level athletic ability have always existed, but now the sporting infrastructure to support training and professionalize female athletes exists, and we have only seen a glimpse of what female athletes will become in the foreseeable future. The progression of more female athletes like Brittney Griner will only put pressure on all sporting cultures to rethink gendered- and sex-differentiated competitions. But as the evolving realities of Caster Semenya's life indicate, global sport is not there yet.

Semenya finally won, in convincing fashion, an 800 m gold medal at the Olympics in the summer of 2016, but the disquiet about her right to compete still remained. In the weeks before the Olympics, British marathon world record holder Paula Radcliffe would weigh in on Semenya's place in the games. On July, 21, Radcliffe—channeling Michael Johnson's concerns about the running of Oscar Pistorius—indirectly questioned Semenya's body by surreptitiously cloaking her gender conservatism in the incontrovertible universals of ethics and fairness. Radcliffe contended that, since Semenya was such an overwhelming favorite in the 800 m, the race was "no longer sport, it's no longer an open race."[70] She implied that Semenya had the potential to win more convincingly than Usain Bolt (though in hindsight, she may have considered rethinking the comparison, given Bolt's dominating performances in Rio). This is shocking to hear from an athlete who in her prime was considered unbeatable. In actuality, as journalist Jeré Longman noted, "Radcliffe is more of an outlier than Semenya. Radcliffe's marathon record of 2 hours 15 minutes 25 seconds is about 10 percent slower than the fastest men's time of 2:02:57. Meanwhile, Semenya's best performance at 800 meters of 1 minute 55.33 seconds, which is not the world record, is about 12 percent slower than the men's record of 1:40.91."[71] Radcliffe used her issues with Semenya as a venue to express her fear that countries would seek out intersexed athletes and female athletes with hyperandrogenism to undermine her version of sporting purity. She opined that "the ethics of fair sport, of fair play, are being manipulated."[72] Nevertheless, Radcliffe is misguided that disciplining non-normative bodies will get us any closer to ethical or fair competition.

Currently, gender verification testing is presented as a way to maintain equality and fairness, but is it really about fairness? Could this testing serve another more relevant purpose? So much about contemporary testing is about setting limits, but is that what sport is really about? Yes, games have rules, but the rules of play seem more flexible than those for the bodies that compete. This is strange, because sport has traditionally been one locale where "unnatural" bodies have transformed the nature of play. Nevertheless, as even the most casual grade school kickball participant knows, the world of sport is far from fair and equal. Sport is about creating, exploiting, and maintaining inequalities. It is really about how to garner and exploit the largest legal or illegal competitive advantage possible. Guy Wilson-Roberts eloquently summed up the nature of sport on his blog *Le Grimpeur* when he wrote: "Sports . . . are the backdrop of suffering, sacrifice, joy, heartbreak, greed and deception."[73]

As the physical and athletic void between bodies classified as men and those categorized as women closes, what will it take for sporting competitions to abandon sex in exchange for body-based competitions? First, it would require a significant culture shift in the way the world sees sex and gender. Outside of the relative size and strength issue, one of the larger problems is that women competing against men are a direct confrontation to historically rooted masculine sport cultural narratives, which many want to maintain. Even many scientists who are critical of the ways in which genetic information has been used to determine sex in sporting competitions, and who seem to support the post-gender cyborg athletic body, nonetheless believe that there are "sound reasons for separating the sexes in athletics including history and historical continuity of the sport, safety, and competitiveness."[74] Although research clearly points in a specific direction, many scientists have yet to accept that "the cyborg is a creature in a post-gender world."[75]

By embracing the diversity of bodies outside the contexts of gender and sex, it may be possible to redefine what constitutes a relevant competition. This is not to say that sex- and gender-based competitions should be eliminated, but if comparative bodies, rather than sexes or genders, were to compete against each other, more interesting, closer, and potentially fairer competitions would undoubtedly result. This approach may be a more productive and constructive use of the multiple testing protocols, procedures, and methods used within sport. In our emerging technoscientific era, will publics, competitors, and sport governing bodies be able to detach themselves from the tenets of sex-based competition in the interest of seeing more balanced competitions with equivalent athletes? Feminist critiques of sport can greatly inform this debate and provide "a way out of the maze of dualisms in which we have explained our bodies and our tools to ourselves."[76] Regrettably, restructuring sporting competitions to consider non-gender-based formulations is untenable at present. Sport, and arguably most modern societies, are so deeply invested in sex-differentiated sporting competition that governing bodies and publics willingly ignore the prevailing social and cultural dialogues that continually expose the difficulty of conclusively defining what makes someone male and what makes someone female. Unfortunately, society's trust and belief in gendered technoscientific testing only reinforces these cultural investments.

5

The Parable of a Cancer Jesus

Lance Armstrong and the Failure

of Direct Drug Testing

On June 12, 2012, the United States Anti-Doping Agency formally charged Lance Armstrong, Johan Bruyneel, Pedro Celaya, Luis Garcia del Moral, Michele Ferrari, and Pepe Marti with anti-doping rule violations. The charges purported that all six men provided medical or management support for the United States Postal Service, Discovery Channel, Astana, and RadioShack cycling teams during the years Lance Armstrong was a member. The USADA contended that it possessed evidence proving that from 1996 to 2010, these individuals developed and sustained a doping program to systematically administer the banned substances of erythropoietin, testosterone, human growth hormone, and corticosteroids, as well as perform the illegal practices of transfusing blood, infusing saline, and injecting plasma intravenously.[1]

The USADA letter was not the first time these men had been embroiled in accusations of doping. Questions about their involvement with illegal performance-enhancing substances had circulated around them since the late 1990s.[2] Ferrari had been well known for his ability to "prepare"—a familiar euphemism for programmatic doping—the world's best cyclists, such as Tony Rominger, Mario Cipollini, and, most importantly, Lance Armstrong, to produce amazing athletic performances.[3] Though the USADA charged six men, the focus and reporting centered on Lance Armstrong and once again launched a public sporting debate around him and performance-enhancing drugs.

Armstrong initially responded to the charges with: "I have never doped, and, unlike many of my accusers, I have competed as an endurance athlete for 25 years with no spike in performance, passed more than 500 drug tests, and never failed one."[4] The veracity of these claims was at the center of the ongoing debate. Though many of the USADA's claims now have been substantiated (Armstrong admitted to years of drug use during a celebrity-styled interview with Oprah Winfrey), there are still unresolved questions about how he was able to get away with it for so long, who knew what and when, and, more importantly, why the tests were so ineffective. Public discussion and debates on- and offline about this not-so-new challenge to Lance Armstrong's cycling record, cancer activism, and cultural heroism are instructive sites to understand the social, cultural, and political meanings of direct drug testing within sport.

The USADA's allegation was news to those outside cycling's inner circles, but it was not particularly shocking. Professional cycling's public image, in relation to performance enhancement, has been a bit shaky for quite some time. Ever since the July 8, 1998, arrest of Festina soigneur Willy Voet by French customs officers at the Belgium/French border (he was stopped in a vehicle stockpiled with narcotics [cocaine and heroin], EPO, HGH, corticosteroids, amphetamines, and syringes to dispense the drugs), professional cycling has struggled to reform its reputation as nothing more than the quintessence of a drug-laden sport.[5] This event is an important marker for the story of Lance Armstrong and doping because it shed light on the seedy workings of professional cycling and exposed doping on an international scale.[6] It moved the public perception of doping away from rogue individuals to a place where drug use was seen as systematic, not just within cycling but in all elite sporting communities.

The USADA placed Lance Armstrong on a historical continuum of high-profile and celebrated athletes charged with using banned substances. Whether Lance Armstrong or any other banned athlete used performance-enhancing substances should no longer be a central question evaluated by multiple sport governing bodies, athletes, and publics. In reality, these constituencies should ask deeper and more substantive questions of themselves, interrogating why all forms of testing have been so ineffective in curtailing drug use, what institutional functions these tests serve, and how the resulting data is deployed to quell and mollify public concerns about the fairness within sport.

The Armstrong Effect

Sport governing bodies—specifically the USADA, World Anti-Doping Agency, International Olympic Committee, and Union Cycliste Internationale—spent a great deal of time arguing over which group possessed the jurisdiction to determine if Armstrong and his charged compatriots should receive sanctions. These debates were significant because these organizations did not agree on how and if sanctioning Armstrong and the five others charged would benefit cycling and sport in general. Those in favor of sanctioning, such as the USADA, believed that it would be an important step in cleaning up not just cycling but sport at-large. The punishment of Armstrong would send a forceful message to professional cyclists, amateur cyclists, and the sports world in general, making it clear that the USADA had the will, power, and authority to police sport in order to keep the playing field fair and equal for all. The UCI asserted that attacking one of cycling's greatest heroes would only dredge up past indiscretions of a troubled time that many hoped to forget and would not allow cycling to move forward from an unpleasant past. The situation became even stranger when the WADA and the UCI began to weigh in on the jurisdiction issue.

The WADA believed that the USADA had the right to sanction Armstrong as well as the others charged. However, the UCI contended that it should have final responsibility to decide if Armstrong and the others should receive sanctions. USADA attorney William Bock, in a reply to the UCI's statement contending that it should be responsible for overseeing the case against Lance Armstrong and others, wrote: "By our count, of the twenty-one . . . podium finishers at the Tour de France during the period from 1999–2005 only a single rider other than Mr. Armstrong was not implicated in doping by a subsequent investigation. Yet, only a single one of these riders had a positive test with the UCI. The rest of the podium finishers were implicated by law enforcement investigations."[7]

The UCI's position was particularly troubling because it had traditionally supported the rulings of national governing bodies and often had been highly critical of national governing organizations who proffered light penalties for athletes caught using banned substances. Part of the UCI's position most likely was driven by its concerns for its own public image. The case would eventually show that the UCI and its leadership problematically intertwined its desire to increase the global reach of its sport with Armstrong's ascending career and actively turned a blind eye to his infractions so as not to undermine their sport and its most marketable athlete. In this regard, the UCI ignored, then denied, statements by Armstrong's former teammates Floyd Landis and Tyler Hamilton that he failed a drug test during the 2001 Tour de Suisse and that the UCI covered it up.[8] But the UCI had a less-convincing response when news reports in 2010 revealed that Armstrong and his management company, Capital Sports and Entertainment, had donated $25,000 and $100,000 to the UCI for anti-doping activities in 2002 and 2005, respectively. The UCI contended that these donations were not untoward and that it had nothing to hide regarding its relationship with Lance Armstrong, whereas multiple reports viewed these donations as highly unusual at best.[9]

As the Armstrong case drew more public and media attention, largely from the news feeding frenzy driven by the salacious pleasure of watching a heroic legacy unravel, deafening silence characterized the comments from the professional cycling peloton. Armstrong possessed a great deal of power, and most in the world of competitive cycling were unwilling to potentially out themselves as knowledgeable of or involved in Armstrong's, or any other, illegal activity by saying anything about the USADA's charges. But more importantly, those in professional cycling did not want to transgress the sport's omertà—the collective secrecy regarding illegal activities in cycling, ranging from doping to paying competitors to "throw" races.[10] Lance Armstrong was known as a vicious defender of omertà.[11]

As for the larger public's perception prior to Armstrong revealing his usage, it was difficult to see any significant movement in the two poles of public opinion. On one extreme was the belief that Armstrong participated in cycling during a historical moment in which performance-enhancing drugs were a necessary evil. If he wanted to reach the pinnacle of the sport, he had to knowingly cheat and use the compounds, products, and procedures necessary to compete for victory. This sporting environment necessitated that Armstrong actively deceive invested publics and mislead scrutinizing governing organization by lying about his dedicated drug usage. To benefit his cycling ambitions and defend against the potential disclosure of his secrets, this sporting ecosystem also demanded that Armstrong also enroll his teammates in his systematic drug program.

The other end of the spectrum held that Armstrong was the greatest cyclist of his generation, if not ever. His recovery from cancer was heroic and inspirational. He had never tested positive for a banned substance, and those who would imply that he used performance-enhancing drugs were just trying to unfairly take him down because they were jealous, had larger political motives, or were doing it for their own personal gain.

As the case broke in the news, Armstrong was so polarizing that there was very little space, discussion, or debate between these two overbearing extremes. This is where testing caught a snag, because it was incapable of providing conclusive answers. It had very little power to unravel and alter deep-seated personal convictions. On the pro-Armstrong end of the spectrum, no technoscientifically based evidence would be persuasive enough to overcome Armstrong's declarations of purity. On the anti-Armstrong end, Armstrong's proclamations that he did not use performance-enhancing substances could not vanquish the overwhelming cloud of suspicion and the belief that technoscientific testing would prove him wrong. Both perspectives have compelling histories behind them. Our society is replete with instances in which the most trusted individuals consciously deceive to maintain privilege on the one hand and public agencies falsify data to preserve power on the other.[12] Furthermore, testing as an evaluative tool for social and cultural decision making—whether in education or in sport—has fallen short of yielding the truth-making explanatory power that is unfairly foisted on it.[13]

For Armstrong, it would not be the smoking gun of positive drug tests that would crumble his sporting, public, and private lives; it would be the legal pressure put on his fragile personal network that would lead to his downfall. As effective as governing bodies have hoped testing technologies to be, they frequently fail in their ability to influence and shape the social and cultural aspects regarding how the public responds to or feels about their athletic heroes. Thus in many instances the battle was not as much about the usefulness and merits of testing, but more about individual and collective public opinion and belief. Since all types of testing data and metrics struggle with the issues of discernibility, the facts do not necessarily speak for themselves. They need to be translated by a trusted and valued interpreter. Throughout the pursuit of Armstrong, it would appear that this reality was something that the USADA understood much less than Lance Armstrong's legal team.

Cleverly, on July 9, 2012, in response to the charges of June 12, Armstrong sued to block the USADA from proceeding with its case and denounced the USADA and its CEO, Travis Tygart, as misusing taxpayer funds to subsidize a spiteful manhunt to take him down. Armstrong's legal defense mirrored his public rhetoric and attempted to shape the narrative by casting the USADA as a vengeful, overreaching, and overspending arm of the government that was meddling in things it knew nothing about. The following day, US District Judge Sam Sparks dismissed this lawsuit and sharply reprimanded Armstrong's legal team with the following statement: "This Court is not inclined to indulge Armstrong's desire for publicity, self-aggrandizement or vilification of Defendants, by sifting through eighty mostly unnecessary pages in search of the few kernels of factual material relevant to his claims." Sparks concluded by asserting that, "contrary to Armstrong's apparent belief, pleadings filed in the United States District Courts are not press releases, internet blogs, or pieces of investigative journalism. All parties, and their lawyers, are expected to comply with the rules of this Court, and face potential sanctions if they do not."[14] Though Sparks admonished Armstrong's legal team, he did allow them to refile a more legally relevant suit within twenty days. Tygart and the USADA claimed that they did not care about public opinion. But this was naive because part of the Armstrong defense hinged on dragging them, whether they liked it or not, into a very public dispute.

Questions about taxpayer support of the USADA gained political vigor with a July 13 letter from Wisconsin Congressman James Sensenbrenner to the Office of National Drug Control Policy (ONDCP) calling for explicit clarification on the ONDCP's "distribution of $9 million worth of funding allocated to the United States Anti-Doping Agency" and insisting that the ONDCP do a better job of overseeing the USADA.[15] Within the week, Lance Armstrong Foundation lobbyists were speaking with José Serrano, a member of the House Appropriations Committee, regarding the "fairness" of the USADA and its actions against their foundation's namesake.[16] The political stakes rose on July 13 when United States Senator John McCain extended support to the USADA and validated "its right to undertake the investigation of, and bring charges against, Lance Armstrong."[17] This congressional interest did not subside and even reached down to the state level when California Democrat State Senator Michael Rubio and twenty-two other state legislators requested that California

Senators Barbara Boxer and Dianne Feinstein investigate the USADA.[18] Thus the investigation of Lance Armstrong was significantly different than that of any other athlete the USADA had attempted to sanction. Armstrong possessed the political heft to turn his investigation into a federal concern.

The refiled case appeared before Judge Sparks on August 17, and he again decided on behalf of the USADA on August 20. Judge Sparks based his final dismissal on the fact that "Armstrong's due process claims lack merit" and that "the Court lacks jurisdiction over Armstrong's remaining claims, or alternatively declines to grant equitable relief on those claims."[19] Armstrong now had a few days to decide whether to accept the USADA sanctions or to proceed with arbitration. On August 24, Armstrong, in his familiar embittered tone, agreed to no longer contest the charges. Arguing that there was no way for him to receive a fair trial, he declared it was time to quit and move on with his life.[20]

The heart of this legal battle pivoted on whether Armstrong and others under investigation created, directed, and participated in years of systematic doping and test evasion. Statistically, the numbers were not in Armstrong's favor, because since Tom Simpson's dramatic death on Mont Ventoux during the 1967 Tour de France, "86% of Tour de France winners have been sanctioned for or incriminated in doping activity."[21] This extremely high percentage makes a strong case for the history of illicit drug activity in professional cycling. Furthermore, the number of athletes from all sports admitting to using a banned substance without testing positive grows longer as each year passes.[22] These facts did not bode well for Armstrong and his compatriots.

Public knowledge of drug use in sport was far from a new phenomenon when the USADA brought its case against the Armstrong group. Amphetamines and anabolic steroids were two classes of drugs that propelled the IOC to create a medical commission in 1967 to investigate and curtail the use of performance-enhancing substances.[23] The cases of East Germany's process of institutionalizing doping, Victor Conte and the Bay Area Laboratory Co-operative (BALCO) administering doping agents to a host of American athletes, and the WADA's report outlining state-sponsored doping of Russian athletes at the 2014 Sochi Olympics highlight the skill and sophistication that athletes, scientists, and nation-states undertake to evade the testing units of athletic governing bodies in order to use performance-enhancing substances, all in the name of winning.[24] Situated within this history, one had to work very hard to avoid reaching the conclusion that Lance Armstrong was not a key player in a doping network, regardless of whether he passed on the opportunity to take the USADA case to arbitration.

Since the case did not go to arbitration, those who were not privy to evidentiary documents possessed by Armstrong's legal team or the USADA were left to settle on one of two conclusions: (1) that Armstrong avoided arbitration because the USADA rigged the process against him so as to guarantee it would win, or (2) that the USADA possessed such damning evidence through witness testimony that Armstrong would only further tarnish his heroic image by allowing this evidence to be seen by an international public. These options left Armstrong in the same position that he was in before the USADA brought charges against him. But, more importantly, the outcome did not shed any new light on the use and efficacy of old or new testing instruments. Since he declined arbitration, no evidence was available to debate. Only the history of speculation, inference, and innuendo was available for supporters and critics to evaluate. Even though the case centered on Armstrong and his potential drug usage, from a technoscience and sport perspective it was also about the public's understanding, belief in, and acceptance of the efficacy of drug testing. In a sense, it was not only the Armstrong group but also drug testing itself that was on trial.

In the end, Armstrong admitted to years of systematically using illegal forms of performance enhancement, but the admission did not come on the heels of a positive drug test. Therefore, his incessant claims that he had never tested positive for performance-enhancing substances are mostly, if not completely, true. Thus did his admission change anything about the perception of testing for performance-enhancing substances? In thinking about Armstrong's impact on sport and the politics of testing for performance-enhancing substances, it is important to ask the following questions: (1) what is the current meaning and value of testing within sport? and (2) is direct testing an appropriate or useful way of regulating the use of banned performance-enhancing substances in sport? The events around Armstrong present a useful location to explore both of these questions.

The Making of a Cancer Jesus

Before Armstrong and other high-profile athletes, such as Marion Jones, admitted to using performance-enhancing substances, testing could support multiple cultural narratives about sport. For instance, one could argue that

1. sports such as cycling and track and field are mostly "clean" because governing bodies such as the UCI and IAAF administer forms of testing that catch and deter cheating by athletes;
2. cycling and track and field are inherently "dirty" sports because so many of the leading athletes have been sanctioned;
3. cycling and track and field are inherently dirty because multiple tests have proved that athletes regularly use banned substances; or
4. cycling and track and field are clean because so few athletes have ever failed a doping test.

Sports such as American football and baseball can sidestep these discussions because they have had such weak testing protocols historically.[25] The interesting part is that the testing allows us to happily and unquestioningly consume the narrative of our choice. Not surprisingly, these dominant narratives require a simple reading of testing.

A more careful analysis of testing destabilizes the power of these valued and comforting narratives. For instance, in the case of Lance Armstrong, it is important to note that during the majority of his cycling career no test existed to expose the use of the two most potent forms of performance enhancement—blood boosting and EPO. Though it may be forever difficult to ascertain the specific details of whether positive tests for EPO and corticosteroids were brushed aside by the UCI during Armstrong's career, it points to the core of the ways in which Armstrong and other elite athletes of his generation maintained their unadulterated public images prior to admitting drug usage.[26]

In many ways, Armstrong's reality mirrors that of sprinter Marion Jones. It was only after her tearful admission to using performance-enhancing drugs that the narrative about her innocence evaporated. Through many athletes' careers, they rely on the fact that no fully confirmed direct test has proven that he or she has used a banned substance. These athletes appeal to fans, hush critics, and manage a host of different publics on these grounds. Often the appeal is so intoxicating, or, in the case of Armstrong, so intimately married to a godlike status of a cancer survivor, that it is difficult for critics to overcome supporters' emotional connections to a heroic story and enter into a critical dialogue about the limitations of tests and testing results.

Buzz Bissinger's August 27, 2012, *Newsweek* commentary "I Still Believe in Lance Armstrong" exemplifies this dilemma. He unabashedly declared that Lance Armstrong "is a hero, one of the few we have left in a country virtually bereft of them. And he needs to remain one." It is hard to disentangle Armstrong's heroism from American heroism, and Bissinger has every intention of keeping them grafted together. But more disconcerting is his willingness to dismiss any concern over Armstrong's potential drug use. "Did he use enhancers? Maybe I am the one who is blind, but I take him at his word and don't believe it; he still passed hundreds of drug tests, many of them given randomly. But even if he did take enhancers, so what?" It is Bissinger's flippant "so what?" that confirms the limited power of testing to provoke significant social or cultural change on this issue. But it also alludes to the nationalist, racial, and masculinist rhetoric around Armstrong. Bissinger, along with his supporters, saw Armstrong as an exemplar of American excellence that must be rallied around, supported, and continually uplifted. The narrative arc of the Armstrong story strongly references a modern-day American "Great White Hope," exemplified in Howard Sackler's 1967 play of the same name.

In the United States, the dominant narrative constructed Armstrong as the greatest cyclist ever. Though most of the world agrees that the greatest ever is unquestionably Eddy Merckx, Armstrong's narrative demanded that he become the greatest. It was not just any heroism, but it was decidedly an American heroism. Part of the allure was that in an age of shifting concentrations of global power, Armstrong's perceived domination of a distinctly non-American sport—though by the end of his career, he only really rode the three-week Tour de France out of months of potential races—confirmed that the United States was *still* a great country. Former teammate and eventual critic Tyler Hamilton precisely captured this American heroism when he called Armstrong "the headstrong American cowboy storming the castle walls of European cycling"[27] (figure 5.1).

Figure 5.1. Lance Armstrong winning the first mountain stage, to Sestrières, Italy, in the 1999 Tour de France. His astonishing performance against the best climbers in the world renewed suspicions about doping usage, which eventually unraveled his cycling career and triumphs. REUTERS / Alamy Stock Photo

Armstrong represented a special type of American hero. He bolstered a collective confidence within the United States' fragile hold on nostalgic-laden perceptions of global financial, political, and military domination and control. Armstrong and his nationalistic hubris reinforced this narrative. The bold airliner attack on the twin towers of the World Trade Center in New York City shattered an American sense of dominance and invulnerability. Armstrong's heroic image was cast in the shadow of a renewed call for American exceptionalism following the attacks.[28] He represented one version of all that was supposed to be good and pure about American society. But, most importantly, Armstrong, through his cycling exploits, showed that America could overcome tall odds and beat the world's best at their own game, and on their own turf.

Armstrong's cancer survival only added a mythical, and arguably godlike, quality to the heroic American narrative. The publication *It's Not About the Bike: My Journey Back to Life*, authored by Armstrong and Sally Jenkins, created and fortified the image of Armstrong as a cancer Jesus.[29] Armstrong vigorously promoted and wielded this fabricated image. While being interviewed after his seventh Tour de France victory in 2005, Armstrong aggressively retorted: "The people who don't believe in cycling, the cynics, the skeptics, I feel sorry for you. I'm sorry you can't dream big and I'm sorry you don't believe in miracles."[30] This spontaneously crafted statement, and many others like it, sadistically melded narrative elements of American exceptionalism, Manifest Destiny, religious resurrection, and the American Dream to produce an irresistibly seductive narrative mélange. This rhetorical amalgam became so potent that it implored Armstrong's many supporters, such as Bissinger, to willingly overlook and repeatedly disavow the expanding fissures in his narrative in order to hold tightly onto their unadulterated hero.

One of Armstrong's most ardent critics, Paul Kimmage, noted: "Cancer had made him untouchable. It was his weapon and his shield."[31] Questioning Armstrong or accusing him of lying was an accusation against a cherished and valorized American way of life. Many Americans rallied around Armstrong to defend their country's honor, endorse an American version of white male masculinity, and maintain the illusion that the United States was the lone cultural and political power to which all other countries needed to acquiesce. At times, no level of proof seemed able to crack the fortress protecting this narrative. Positive direct testing evidence had traditionally been seen as the marker of wrongdoing, but Armstrong's case showed that this might no longer be true. It is an athlete's public confession to illegal behavior that has now become the bar.

This testing reality gets to the issue of what level of testing proof is conclusive enough to influence the views of the most skeptical naysayers or ardent supporters. This, of course, depends and varies on the values of a sport culture, the technoscience involved, and the individual and collective desires to be open to information that may force the reevaluation of a cherished sporting competition or athlete. It is generally accepted that a statistically significant result of a technoscientific test occurs if there is a one-in-twenty chance of a false positive result. It is not clear if the general public feels the same way. Often, one in twenty does not reach the level of accuracy that the public demands. Clearly, the tension between scientific levels of accuracy and public levels of accuracy is fraught with complications that science has unquestionably helped create. In this regard, what would it mean if the chance of a false positive was pushed to a higher standard of, say, one in one thousand?

This ratio may be more convincing and able to withstand a higher level of critique. But this level of verification is probabilistically difficult to achieve through current drug-testing regimes. To reach a one-in-one-thousand chance, an athlete would have to be tested approximately 693 times.[32] Very few, if any, athletes have been tested that many times during their careers. Lance Armstrong stated on numerous occasions that he has been tested more than 500 times. That number has been thrown around liberally, but questions have been raised regarding whether 500 tests were actually performed. In fact, the UCI and the USADA confirmed that they only tested Armstrong 215 and fewer than 60 times, respectively.[33] Nevertheless, athletic drug testing is still extremely messy on a large scale because it not only needs to be scientifically valid but also understandable, readable, and convincing to a general public. As difficult as it is to reach irrefutable levels of certainty, the limited number of times athletes are tested and the processes by which athletes dope make it exceedingly difficult to detect banned substances.

The current doping environment may be partially driven by the use of smaller amounts of doping products more frequently to achieve the same results as using larger doses of doping products less regularly. This process, called microdosing, has the added benefit of allowing an athlete's body to excrete the traces of banned substances quickly and stay below levels of recognition. To test the effectiveness of microdosing and EPO, journalist Mark Daly purchased the drugs and began an informal doping program for an episode of the BBC program *Panorama* titled "Catch Me if You Can," aired in 2015. Daly, a fit amateur athlete, showed the effectiveness of the drugs but also exposed how easy it was for him to evade testing positive.[34] In response to Daly's report, the WADA noted that "the programme also [raised] questions regarding the ability of athletes to dope by taking minimal amounts of performance enhancing substances without testing positive, otherwise known as 'micro-dosing'. It is an issue that we are exploring in great detail with experts from across the anti-doping community."[35] Since Daly's blood was not formally tested by the WADA, though he stated that he used the same labs that the WADA uses, it defended itself by concluding that "while the programme suggests that the journalist, through his experiment, was able to enhance his performance without recording an adverse analytical finding (AAF), we haven't been provided any information that would validate this allegation."[36] This brief statement did little to diminish the power of the broadcast. Daly's report highlighted clear flaws within testing but also reinforced the assumption that a positive drug test is an indication that an athlete followed a very poorly constructed doping program. The vernacular is that by testing positive, an athlete failed two tests: (1) a drug test and (2) an intelligence test.[37]

The processes by which athletes avoided testing positive were quite simple. BALCO's Victor Conte, in his letter to British sprinter Dwain Chambers, outlined the highly successful "duck-and-dodge" technique of supplying misinformation to testing authorities in order to "circumvent the British and IAAF anti-doping tests for an extended period of time."[38] He indicated that athletes could miss two tests within an eighteen-month period without sanction, which allowed for a long period of systematic drug use. The letter also chastised the USADA for its lax testing methods during a period when most athletes were duckin' and divin' and using anabolic steroids and other drugs. He noted that "if you check the testing statistics on the USADA website, you will find that the number of out-of-competition drug tests performed during each quarter of 2007 are as follows: in the first quarter there were 1208, second quarter 1295, third quarter 1141 and in the fourth quarter there were only 642."[39] Though intimately involved in providing athletes with illegal performance-enhancing substances, he made a recommendation to the USADA in 2003 that they test more in the fourth quarter of each year.

Conte confirmed that the USADA followed his advice until 2007, at which point it cut back on testing. He implied that the USADA cut back because it wanted to avoid catching too many American athletes before the upcoming 2008 Summer Olympic Games, or as he put it: "This is equivalent to a fisherman knowing that the fish are ready to bite and then consciously deciding that it is time to reel in his line and hook, lean his fishing pole up against a tree and take a nap."[40] Conte's insider view projects a high level of confidence honed through his years of experience assisting athletes in their efforts to avoid positive test results. Conte devised an entire system that schooled athletes in the process by which they could "pass" as non-illegal drug-using athletes. Even when authorities caught athletes with banned substances within their bodies, they applied the tried-and-true techniques of denying and attacking.

When a test indicates that an athlete has diverged from a sporting competition's rules and regulations, most athletes' first response is to displace blame and dispute the test results. The basic idea of a false-positive argument is to claim that an error, human or technoscientific, was made at some point during or after the taking of the sample in question. This approach calls into question everyone who interacted with the athlete or the sample; the error could have been made by a trainer giving an athlete a supplement with a hidden banned substance, or the sample could have been mishandled by someone performing the test, or by a machine.[41] Regardless, the athlete disavows all responsibility. In a world where humans make mistakes all the time, it is not out of the realm of possibility that someone made an error. This approach not only locates blame within human ineptitude but concomitantly casts aspersions on the technoscience undergirding the testing.

Travis Tygart credits Marion Jones's public relations team for pioneering this approach. Jones effectively deployed these tactics throughout her battle against the USADA. For a while, these diversionary maneuvers succeeded in redirecting the focus from the data procured by the USADA. In an interview with *ESPN the Magazine*, Tygart noted that "Jones drafted that playbook, and it's been handed down to athlete after athlete, particularly in high-profile cases." In light of Armstrong's vociferous denials, Tygart indicated, "Armstrong's PR team was the same one Marion used."[42]

The drugs and methods of evasion created a seamless web of techniques to avoid being seen or perceived as using illegal performance-enhancing substances. This form of "technoscientific passing" not only facilitates the presentation of "an altered external identity, but also requires the technological user to agree to a sort of temporary amnesia" that he or she is not doping.[43] It is the collective amnesia of athletes, governing bodies, and fans alike that allows athletes to pass as pure, unadulterated athletic champions in instances where this is simply not the case. But it also allows us to avoid questioning the roles of technoscience, and specifically technoscientific testing, in constructing and maintaining heroic narratives, because we as a society have invested in the truth-giving potential of science and technology.

The emotional, financial, and cultural tolls on athletes, fans, governing bodies, and sport in general before, during, and after the pursuit of highly esteemed athletes such as Lance Armstrong, Marion Jones, and Barry Bonds demand that we reconsider seriously what functions the institutionalization of anti-doping testing regimes serve, as well as the ways in which our trust in the technoscientific efficacy of direct testing supports and gives power to these bureaucratic systems. Most simply, testing functions as a technoscientific instrument to provide a way of ensuring that athletic competitions are fair and that each competitor has an equal chance of winning. Thus testing, at its core, is designed to make sure that no athlete has taken a substance or altered his or her body in a way that provides a significant advantage over his or her competitors. Unfortunately, this simple function lacks robust explanatory power.

Testing within sport serves many purposes, needs, and desires. For sport governing bodies, testing does serve the equal competition goal, but sport governing bodies also have an important financial motivation to test. The economic stakes emanate from a sport governing body's institutional purpose of promoting athletes and athletic competitions that are collectively understood by public audiences as fair and just. Testing has become an important and powerful way to establish and sustain the perception that fairness is maintained within sporting competitions. This task has become increasingly difficult to manage because various publics do not conceptualize fairness or inequality in the same way. Nevertheless, sport governing bodies must do their best to produce sporting competitions that address multiple versions of fairness. What is at stake for the sport governing bodies is their existence. If consuming publics do not believe that sporting competitions are fair, or at least buy into an illusion of fairness, they may not watch and, subsequently, not provide a given sport with the necessary financial support to sustain it. Similarly, athletes have to believe that they are competing on a relatively even playing field, and testing, as well as the fear of suspension or sanction, at minimum provides a set of conventions to which all athletes must conform. Thus sport governing bodies, publics, and athletes must buy into the testing regime of a sport.

Governing bodies, for the most part, also oversee sport and create, develop, and maintain the rules that determine how a sport is played, who can play, and what equipment can and cannot be used. Formula 1 automobile racing is named after the "formula," or the rules put into place after World War II to set parameters for engine size, vehicle dimensions, and various equipment. The sport in its contemporary incarnation continually tweaks its "formula."[44] Athletes, publics, and technoscientific actors justly and unfairly criticize sport governing bodies for concocting arbitrary rules that negatively impact a given competition and a way or style of play. New rules or amendments to existing rules will always be a point of contention within any sporting competition, and the rules around testing continue to be a flashpoint because they are some of the most invasive. Specifically, to be in compliance with "WADA's 'whereabouts' rule, all athletes must make themselves available to drug testers for one hour a day, between 6 a.m. and 11 p.m., ninety days in advance, for out-of-competition testing."[45] For some athletes, this steps beyond governing sport and into the realm of onerous invasions of privacy.

Testing also provides an effective way for sport governing bodies to outsource fairness, or, at minimum, share the responsibility for balanced competitions. By relying on the authorial power of technoscience imbedded within testing infrastructures to legitimate themselves and their practices, sport governing bodies can side step the direct responsibility of determining which athletes or athletic bodies can and cannot compete. This responsibility can be, in many instances, subcontracted to those administering the tests. Though governing bodies sanction athletes, they make their ruling decisions based on test results performed in laboratories. To avoid the perception of conflicts of interest, most governing bodies have tests performed by partner labs over which they have no control. Even as sport governing bodies use these laboratories to move the act of testing farther away from their organizational centers, they are still the makers, keepers, and enforcers of the rules that define the boundaries of a sport and its competitions.

The authority of testing as a system rests on its reliance on science and technology. We live in a world where we, in general, are comfortable with the technoscience that mediates our everyday existences. It is understood that science and technology can fail, but for the most part citizens of the world believe that science and technology are necessary to living on Earth. Sport has drawn on this progressive technoscientific history and deploys testing—now rooted in scientific procedures and technological machinery that most do not understand—as the fairest method to determine which athletes can and should compete within a given sporting competition. As long as athletes conform to the testing requirements of a specific sporting competition, the power of science and technology goes unquestioned.

The science and technology of testing are infallible until someone tests positive, at which point attempts are made to redefine science and technology from rational activities to indiscriminate social practices. Depending on the athlete, positive tests have also brought the public into the discussion, and in these debates the specifics of test results often do not matter much. For certain athletes beloved by their followers, such as Lance Armstrong or Marion Jones, no amount of testing evidence could have been powerful enough to change followers' opinions that someone undoubtedly made a mistake. This reality is the crux of the tension around testing. Most athletes who have used banned substances rarely test positive more than once. This is because once a positive test has been confirmed, the athlete is usually banned, sanctioned, ostracized, or, in rare cases, has launched an effective defense. But because testing for banned substances is limited to known agents and the technoscience behind evasion is so strong, very few athletes go through this process at all. This reality flies in the face of those who contend that testing has been effective. It subsequently has led organizations such as the USADA to rely on the testimony of close confidants to determine if an athlete transgresses a sport's rules and regulations.

The Undoing of a Cancer Jesus

The last act of Armstrong's testing saga commenced on October 10, 2012, when the USADA released its *Reasoned Decision*. By all accounts, the 202-page report, summarizing the more than 1,000 pages of documents submitted to the UCI, delivered on its promise to provide overwhelming evidence. But it was not the technoscientific testing evidence that was the most persuasive. Christopher J. Gore, the well-respected head of physiology at the Australian Institute of Sport, analyzed the blood data and concluded that

Armstrong's disturbingly "low reticulocyte percentage during the 2009 and 2010 Tours de France, coupled with his unusual decrease in calculated plasma volume during the middle of the 2009 Tour de France, build a compelling argument consistent with blood doping."[46] The report buried Gore's cogent scientific assessment 140 pages deep into the text.

This statement appeared so late in the text because in this case, as in many recent drug testing cases, expert analysis, though very important in framing the overall argument, does not necessarily convince nontechnical audiences. In order to determine guilt or innocence, general publics have been highly uninterested in sorting through detailed scientific studies and acquiring the necessary knowledge to understand the meaning behind Gore's description of the ways in which reticulocytes and blood plasma, and one's blood in general, react to the infusion of older red blood cells. When celebrated champions are accused of testing positive for banned substances, interested publics really only want to know if the accusations are true or false.

In Armstrong's case, the clear lack of a failed drug test, which was the type and form of evidence that the anti-doping movement promised as the gold standard of proof for years, was not present. Cycling journalist Neal Rogers summarized this drug-testing conundrum best when he wrote: "As for the remaining *Armstrong faithful*, at this point, it's unlikely they will be swayed. Because of his amazing cancer story, Armstrong is one of the few athletes, or even celebrities, who inspires absolute blind faith."[47] Rogers continued to expose the problems with the lack of a direct positive test and inveighed that "those undeterred by Landis' accusations, Hamilton's book, or the fact that Armstrong opted out of an arbitration hearing, aren't going to be convinced based on the evidence in the case file. In the face of anything less than an admission, they will likely remain loyal to Armstrong . . . and the rest of the Armstrong brand."[48]

It was this belief in the Cancer Jesus that the *Reasoned Decision* aimed to unravel by exposing how Armstrong led a "massive team doping scheme, more extensive than any previously revealed in professional sports history."[49] These are strong words, but they were used to make the indelible point that a significant number of people supported and sustained this doping network. Though the documents included Lance Armstrong's questionable blood data and records of more than a million dollars in payments to Dr. Michele Ferrari—the well-known advocate for and specialist in the administration of performance-enhancing substances—the most convincing aspects of the dossier for the USADA were the damning testimonies given by eleven of Armstrong's closest teammates, detailing the minutest details of the sordid drug business Lance Armstrong orchestrated. Specifically, the statements by highly regarded cyclists George Hincapie, Levi Leipheimer, and Christian Vande Velde—none of whom had ever tested positive for any banned substance or admitted drug use—were the most persuasive aspects of the USADA case, but they also effectively reshaped the general public's perception of Lance Armstrong. The heft of these collective commentaries outweighed the efficacy of any testing evidence examined by Christopher Gore.

The questioning of testing efficacy by numerous fans, journalists, athletes, and scientists weakened the potency of this information so much so that an organization such as the USADA, in Armstrong's case, relied more heavily on the direct personal expert testimony of Armstrong's former teammates than amorphous direct testing evidence or indiscernible expert scientific analysis. The USADA, and the anti-doping movement in general, used Armstrong as an opportunity to reconfigure the terms of how we all judged infractions of a sport's rules regarding doping. The anti-doping movement has demanded unwavering faith in the direct testing model, which for Armstrong's most ardent supporters makes the *Reason Decision* unconvincing because it lacks this most potent form of evidence. As evidentiary explanatory power migrates away from the technoscience of direct testing, it becomes exceedingly more difficult to explain the nuances of new and emerging indirect methods and convince multiple publics of their efficacy in light of the anti-doping movement's historical claims that direct testing is reliable, definitive, and incontrovertible.

The tragic fallout of Armstrong's doping moved beyond the world of cycling and media commentary to his financial well-being, when, on October 17, Nike—historically one of Armstrong's most ardent supporters—ended its contractual relationship with him. Nike's brief press release stated, "Due to the seemingly insurmountable evidence that Lance Armstrong participated in doping and misled Nike for more than a decade, it is with great sadness that we have terminated our contract with him."[50] To further distance itself from the tainted athlete's doping transgressions, the company emphasized that it "does not condone the use of illegal performance-enhancing drugs in any manner."[51] Nike's public distancing from Armstrong came the same day the world found out that Armstrong would relinquish the chairmanship of the Livestrong Foundation. Nike severing ties with Armstrong opened the floodgates, and all his other major sponsors, from Anheuser-Busch to Trek Bicycle Corporation, abandoned him within hours of the Nike press release. Public and consumer opinion on the *Reasoned Decision* was clear when reports surfaced that some wearers of the yellow Livestrong bands began deleting the *V* to create their own customized "Liestrong" bands.[52]

The public and consumer response is fascinatingly tied to Nike's choice to terminate its partnership with Armstrong. It had much more to do with "transferable attributes" than guilt or innocence. To Nike, Armstrong's value resided within his athletic heroism. He was not just any athletic hero, but an American hero of epic proportions—one that had the marketing potential of upper-echelon athletes such as Michael Jordan. It was not just the narrative of cancer survivor to Tour de France champion that made Armstrong special but also the fact that it was understood that he accomplished it all fairly, purely, and against insurmountable odds. This Cancer Jesus narrative supported the illusion that Armstrong succeeded by outworking everyone else, and not by having a body that responded to technoscientific doping better than others. The heroic transferable attributes made him a marketing tour de force. However, the *Reasoned Decision* drastically altered the cognitive reception of Armstrong. The Armstrong narrative quickly descended from "'hope' and 'courage' and 'hard work' . . . to . . . 'deceit,' 'immorality' and 'short cuts.'"[53]

Nike, as a technoscientific actor, the public, and the UCI all invested in Armstrong as a representation of all that was good and pure about sport. With so much fantastical hope and so many unrealistic dreams foisted on him, it is not all that surprising that he betrayed them all in the end. This point is most relevant to Nike, which has been known to give unwavering support to its sponsored athletes when they are in trouble. Nike stood by Kobe Bryant in 2003 after he admitted to a moment of infidelity that erupted in a sexual assault trial.[54] In 2009, Nike supported Tiger Woods during the dismantling of his youthful, "nice guy" image after exposés surrounding his sexual life emerged.[55] It even supported Joe Paterno in 2012 when investigations summarized in the Freeh report concluded that he did more to protect his football program than children from a pedophile on his staff.[56] And though Nike disowned Michael Vick in 2007 for his dog-fighting conviction, it resigned him after his release from prison.[57]

What makes all of their transgressions different from those of Armstrong is that they occurred outside of the competitive athletic arena. Bryant, Woods, Paterno, and Vick, though flawed human beings, did not bring shame to the game by developing a systematic network of cheating and shamelessly lying about it for more than a decade. Thus what makes Armstrong different is that he broke the sporting culture's trust by flagrantly turning his back on the athletic ideal of fair and equal competition. Though this ideal is an illusion, it is one of the foundational premises on which competitive sport is built. Armstrong doped, but not like anyone else. He took it to a higher level—and perfected it. He was always willing to illegally push the technoscientific envelope and demanded his teammates do the same. The systematic coercions frighteningly resembled the way in which pedophiles groom their victims or slave traders seasoned their slaves.[58] Armstrong willingly deployed a powerful technoscientific, social, and cultural network to achieve his goals in a world that wanted to and still would like to believe that unnatural cycling achievements, and for that matter all sporting feats, are achievable *paniagua*—on bread and water.[59]

The UCI begrudgingly confirmed the USADA's sanction on October 22, 2012, and agreed to "disqualify all competitive results achieved by Mr. Armstrong from 1 August 1998."[60] On December 6, 2012, the UCI officially stripped Armstrong of these results.[61] In its decision, the UCI—which many viewed as significantly contributing to the doping problem by protecting Armstrong, other valued cyclists, and its brand by turning a blind eye to the entire business of doping—denied any responsibility for cycling's deplorable state, and maybe they were partially correct in this perspective. The UCI did not control how professional cyclists integrated doping into their collective zeitgeists. Former world hour record holder and world champion Graeme Obree persuasively summarized this period of cycling when asked about doping when he joined the French professional team *Le Groupement* in 1995:

> First time I go to meet the Groupement team, I'm in Belgium and a fellow rider is saying 'Hey, what [drugs] did you use for the hour record?' And I'm like, 'Nothing'. And he's looking me up and down. The guy had no respect for me. It was an Italian guy. 'What did you use', he says again. 'Nothing'. And he goes 'Amatore'. Amateur. I'll never forget that. He turned on his heels and walked away in disgust. I think these

people truly believed that I was being unprofessional by refusing to sign up to the drugs programme. They thought it was proper to take drugs otherwise you weren't taking the sport seriously. Because I wasn't taking the drugs, to them, I was an amateur.[62]

Obree only lasted a year with *Le Groupement*. He subsequently became an outcast when he told the leading French sport newspaper *L'Equipe* that he believed that "99 per cent of elite road cyclists were taking drugs."[63] When then-president of the UCI Hein Verbruggen responded to Obree's statement by calling him a coward, it was unclear if cycling, the UCI, or the anti-doping movement writ large could limit the collateral damage doping caused the sport. As it currently stands, it is still an open question if the sport of cycling will reform itself or continue on as it always has.

For the technoscientific procedures falling under the umbrella of performance-enhancing drug testing, the *Reasoned Decision* revealed troubling limitations in these scientifically, technologically, socially, culturally, and institutionally sanctioned practices. It exposed how testing for performance-enhancing drugs had become hamstrung by its history and terminology. A test is a process by which a form of truth is expected to be determined. It has been constructed as definitive. More importantly, large sections of society accept testing as the only way to reach a form of truth. The windstorm of accusations that circulated around Lance Armstrong in 2012 shows how problematic this is for sport and the desire to make competitions drug-free. If the only way to discern if Armstrong or any other athlete used a banned substance is through direct testing (in order to find an illegal substance within blood or urine), sport has potentially created a testing atmosphere in which the burden of proof is too high for existing testing regimes to reach. The limited effectiveness of current testing also allows the public to believe in and keep their heroes because it is so weak that no positive test can overcome existing concerns about reliability and shake a fan's belief in their heroic champion. It has become exceedingly clear to most that elite athletics is as much about the athletes and their abilities as it is about the legal and illegal material or pharmaceutical technoscience that produces great performances.

Initially, out-of-competition testing was developed in the late 1980s to bolster the efficacy of direct testing, but Victor Conte showed how simple it was to evade these tests. In the past decade, the athlete biological passport has been developed as a new technoscientific instrument of indirect testing designed to test for the residual effects of drug use rather than the specific presence of banned substances. This is potentially a step forward, but the energy around the debate will only increase with the development of new tests that are seemingly less grounded within the scientific practices and technological artifacts with which we have become comfortable. The ABP's increased scrutinization of the body and biological data has the potential to produce more clarity, but it may also make the situation more complicated. Nonanalytical drug and gender testing protocols require science and technology to do something that has always been very difficult. Sport governing bodies, athletes, and fans alike require that these tests provide more clarity on issues that have strong social and cultural resonance. This will increasingly be challenging for science and technology.

Testing, however, will not disappear. Testing is not only about how to simply categorize bodies but also functions as a form of truth making. We live in societies that rely heavily on the outcomes of tests, and for the foreseeable future they will be comfortable and relied-upon forms of analysis. When the normative power of science and technology undergirds these tests, they become all the more authoritative. Since science has become the ultimate form of truth making, scientific testing is constructed as immutable. Within sport, technoscientific testing has been deployed in this manner. But technoscientific testing is just as much a social and cultural process as anything else. It is in this science as truth making that modern sport testing has mutated from evaluations of physical ability and strength to a mechanism for policing the conceptual, material, and physical boundaries of sport. Yes, athletic competitions are physical tests, but the international regime of drug testing dominates the discussion about testing. The public revelations about Lance Armstrong and the sordid business surrounding his systematic doping regime emphatically stripped direct testing from the anti-doping movement as its seemingly most effective form of deterrence and apprehension. In order to more fully understand the social and cultural roles of testing within sport, it has become exceedingly necessary to move beyond simply directly testing for known banned substances to a new space where new and emerging forms of indirect testing will shape the future of sport and the anti-doping movement.

6

"May I See Your Passport?"

The Athlete Biological Passport
as a Technology of Control

The proliferation of performance-enhancing drugs has put professional sport under an uncomfortable microscope. In recent years, Major League Baseball has struggled to manage the public perception of its game in light of drug usage attributed to high-profile players such as Jose Canseco, Mark McGwire, Barry Bonds, Roger Clemens, Alex Rodriguez, and Ryan Braun, to name a few.[1] In the Italian Series A soccer league, the Juventus Football Club has been accused of administering erythropoietin to its players, and questions have persisted regarding the use of performance-enhancing drugs in professional soccer throughout Europe.[2] However, over the past few decades, cycling has become the global centerpiece of the pharmaceutically driven sport. It has become common practice for the press to ponder the purity of every Tour de France champion. The recent confessions of Lance Armstrong and a host of other elite cyclists, along with claims that cycling has cleaned itself up, has done nothing to restore the public's faith in direct drug testing or quell the suspicion of every professional cyclist.[3] The autumnal transition from road cycling to cyclocross seems to never come soon enough for a sport percolating with tensions and accusations.

The Union Cycliste Internationale has been a driving force in the attempt to stabilize the sport and reform its own significantly diminished image. In many ways, this has been the organization's charge from the very beginning. The UCI was formed in Paris on April 14, 1900, when the cycling governing bodies of Belgium, France, Italy, Switzerland, and the United States met in Paris to pool resources and to strengthen cycling as a sport. Beginning with the guidance of its first president, Belgian Emile De Beukelaer, the UCI has effectively managed competitive cycling as it moved beyond its European roots and developed into an international sporting business.[4] Over the past century, the UCI has weathered many storms that have questioned its governing ability. It has attempted to legislate a balance among the advances in material science, engineering, and athletic performance. But its most recent charge, with support from the World Anti-Doping Agency, has been to control the explosion of the most recent generation of performance-enhancing substances.

Initially introduced in the late 1960s, direct testing for specific banned substances has had varying degrees of effectiveness. These tests initially were directed at a variety of stimulants that fell under the category of amphetamines. Developments in chromatography and mass spectroscopy fine-tuned the detection protocols, leading to greater success identifying substances (e.g., diuretics, stimulants, and narcotics) not normally found in the body. Yet over the past several decades, the UCI, the International Olympic Committee, and the WADA have struggled to handle the proliferation of performance-enhancing substances such as recombinant proteins or peptides, which possess the same or very similar chemical structure to those naturally produced in the body.[5]

It has become even more difficult, and in many cases impossible, to create tests for substances that governing bodies do not know exist. Initially, the UCI touted its success in catching offending cyclists as proof that it governed the cleanest professional sport in the world. Though this claim may have been valid, the scale of the investigations and disciplinary actions gave the perception that drug users were not aberrations, but the norm. The UCI's ability to catch and, most importantly, publicize offenders dwarfed all other sports combined. Nevertheless, the UCI contended that if it did not curtail the growing drug problem, its sport would soon be delegitimized completely.

In an effort to regain control of the sport, the UCI redoubled its technoscientific efforts to support the development and creation of a new technoscientific fix. It hoped this new instrument would become the magic bullet that would solve all of its performance-enhancing drug problems. The fix is now known as the athlete biological passport.[6] Yet the ABP, just like all forms of drug testing, is much less about exacting accuracy than about creating a compelling and believable narrative about exacting accuracy. This narrative demands that for direct testing to be useful and meaningful, it has to function as a truth-making mechanism. The supporting narrative for fans, athletes, and all invested in a sporting culture is that direct testing will effectively catch cheaters and deter those considering using banned substances. But the past few decades of sporting history definitively have shown that direct testing has not been particularly successful in this regard.

In 2011, researchers examined the International Association of Athletics Federations' own testing database and concluded that roughly 18 percent of the endurance athletes tested were likely doping in some fashion.[7] The conclusion that nearly one out of every five track and field athletes "cheated" by using performance-enhancing substances is deeply troubling for the narrative that the IAAF oversees equal, fair, and pure athletic competitions. The UCI and sport in general needed a solution, because existing modes of direct testing for banned substances were collapsing under the weight of the UCI's self-constructed narrative of technoscientific accuracy. But this solution, like most technoscientific fixes of this variety, was not only about replacing a less effective test with a more effective one but also fixing the testing narrative to allow viewing and consuming publics to believe, and ultimately trust, that sport governing bodies and athletes were delivering and participating in fair and equal athletic competitions.

So what is this passport? The ABP contains a technoscientifically documented history of an individual athlete's blood and urine profiles. Designed to track a series of biological metrics, the ABP uses longitudinal data from an athlete's blood and urine samples to create a series of "normal" biological markers. What is new and conceivably transformative about the ABP is that it is committed to using an indirect method to detect the use of banned substances. The ABP, as a mechanism for technoscientific surveillance and control, is seen as a means of mandating cleaner sport by the governing bodies legislating its use. From its introduction within the world of professional cycling in January 2008, the ABP gained interest among multiple sport governing bodies.

When the WADA approved its *Operating Guidelines for the ABP* on December 1, 2009, the IAAF, which began collecting and measuring blood values at the 2001 Track and Field World Championships, immediately introduced the ABP.[8] In July 2011, the Fédération Internationale de Natation, swimming's governing body, adopted a biological passport program to combat its own doping problem. In March 2013, the International Tennis Federation followed suit and signed on to a biological passport program. The Fédération Internationale de Football Association used the ABP program for all the players competing in the Men's 2014 World Cup Tournament in Brazil. To fine-tune its processes and procedures, FIFA used the 2013 Confederations Cup tournament in Brazil as an opportunity to gather baseline data on players from the eight countries competing.[9]

To possess a valid biological passport, an athlete must submit a minimum of ten blood and four urine samples to create the requisite baseline profile. In subsequent competitions, if an athlete's blood and urine values veer outside the established limits based on the historical values recorded in the athlete's passport data, the implication is that someone or something has manipulated the athlete's blood or urine. The strength of this approach is that it avoids the need to use direct testing and the detection of a specifically banned compound to determine if an athlete illegally attempted to enhance performance.[10] Instead of chasing every newly developed performance-enhancing substance, the ABP program tracks an athlete's body, through the monitoring of an agreed-upon set of biological markers over time, and utilizes bodies as diagnostic instruments to establish if a banned substance has or has not been used. The ABP aims to have the body police, and potentially betray, any athlete hoping or intending to use banned substances or methods. Prior to the implementation of the ABP, confessions and analytical positives were the most effective ways to determine if an athlete had used a banned substance. An analytical positive occurs when a substance on the WADA's or another sport governing body's banned list is found within the athlete's body after a test.[11] With this model, the burden of proof falls on the anti-doping authority and the techniques used to verify a positive test. Traditionally, a sport governing body is responsible for testing.

The process of determining the use of a banned substance with the ABP is a bit different than with direct testing. Take, for example, cycling and its governing body, the UCI. If the UCI determines that a cyclist's ABP data appears abnormal, it convenes a three-member expert panel. The panel evaluates the data and renders a sanctioning or nonsanctioning recommendation to the UCI. Under the rules of the WADA, the panel has a great deal of flexibility in how it evaluates the passport data. The WADA uses the intentionally hazy term of *comfortable satisfaction* as the level of agreement that a review panel must achieve to recommend the sanctioning of an athlete. The definition of comfortable satisfaction is "greater than a mere balance of probability but less than proof beyond a reasonable doubt." It exists "somewhere between what is normally applied in private law and what is applied under public (penal or criminal) law."[12] The purposeful vagueness of the requirements of guilt conceivably is necessary because review panels come from many different disciplinary and intellectual backgrounds. It would be exceedingly unrealistic to demand that this interdisciplinary group come to some stronger level of agreement on scientific testing data. But it also allows for uncomfortable grayness in decision making. Since the approach lacks the presumed definitiveness of direct testing, it needs to create several gateways of analysis to increase its robustness. The goal is to produce an indirect testing structure that publics believe is even more effective than direct testing.

Even though the burden of proof falls on the anti-doping organization, athletes are responsible nonetheless for making sure that no prohibited substances enter their bodies. The WADA regulations state that "*Athletes* are responsible for any *Prohibited Substances* or its *Metabolites* or *Markers* found to be present in their bodily *Specimen*. Accordingly, it is not necessary that intent, fault, negligence, or knowing *Use* on the *Athlete's* part be demonstrated in order to establish an anti-doping violation."[13] This unwavering language indicates that an athlete is fully responsible for banned substances regardless of how they may have entered that athlete's body. Prior to the institution of the ABP, the list of banned substances propelled all direct testing procedures. That is, substances had to be on a banned substances list in order for a testing protocol to exist. This approach has always left testing agencies and the scientists developing the tests to detect performance-enhancing substances at least one step behind those producing the next versions of sport-altering drugs. Beyond the limitations of athletes using substances that do not appear on the banned substances list, there are other shortcomings to current direct testing methods. Specifically, direct detection requires that an athlete submit a testable sample.

With more sophisticated means of detection avoidance, testing often needs to occur very close to the moment the illegal substances enter the body. Thus if an athlete is not tested during a period of use, he or she will never test positive. Though most sport governing bodies require athletes to keep them apprised of their whereabouts through the Anti-Doping Administration and Management System for regular and out-of-competition testing, the majority of athletes at the elite level are tested only a handful of times each year. These tests tend to occur at or around significant yearly events such as a sport's world championships. Below the elite level, very few athletes are regularly tested for performance-enhancing substances. Anti-doping organizations have been familiar with these limitations from the onset of drug testing. These limitations motivated the critical discussion about creating a technoscientifically based model to perform non-analytic testing represented by the ABP.

The idea of using databases of blood values to determine purity has acquired significant momentum and seems to have shifted the chosen form of performance-enhancing substance enforcement and deterrence from direct testing for known substances to indirect testing for suspect blood analytics. As important, seemingly effective, and widely accepted as the ABP may become, many critics have begun to raise questions about its effectiveness.[14] Though questions about its merits are critically important, the ABP also demands that athletes, governing bodies, and investing fans recalibrate their collective understandings of how illegal performance-enhancing behaviors are tracked, policed, investigated, evaluated, and sanctioned. Part of what the ABP requires is a conceptual reconfiguration of testing for performance-enhancing substances away from direct testing for specific substances to a system that now determines guilt, or innocence, on a set of biological values that in aggregate may provide a more powerful assessment of illegal behaviors. Though the governing institutions supporting the use of the ABP seem to feel that this is a rational step and a simple transition, it arguably may not be that logical or easy for those not intimately connected to anti-doping movements.

In the end, the technoscientific efficacy of the ABP may not be that crucially important because the ABP is as much about limiting illegal behaviors among athletes as it is about creating the perception that a given sport, through the latest technoscientific tools or methods, can produce fair and equal competitions for viewing publics to consume. In this regard, the ABP raises the yet unanswered question, can this or a similar technoscientific tool, device, or method convince a variety of publics to trust the performance of any athlete? That is, does the actual effectiveness of the ABP matter as long as a governing body is able to persuade the public to trust the technoscientific authority of the passport, the legislative strength of a governing body, and the athletes themselves?

Creating the Need for an Athlete Biological Passport

Throughout sport, the mid-1990s saw a precipitous rise in athletes' performances, but the biggest gains took place in endurance sports such as cycling, distance running, and cross-country skiing. Whereas publics marveled at each record-breaking effort, those inside these sports understood that the rapid increases in performance emphatically signaled that athletes had acquired and were using a new and decidedly more powerful category of performance-enhancing substances. In the early 1990s, rumors circulated in the press about endurance athletes such as cyclists experimenting with a new drug that would increase the oxygen-carrying capacity of blood. Though initially unknown, this drug would be revealed as erythropoietin, or EPO. The pharmaceutical company Amgen led the development of EPO. This medically valuable hormonal drug plays a key role in the production of red blood cells. In the life cycle of the drug, it has found important and lifesaving uses in the treatment of chronic anemia, seen in kidney diseases and cancers. It is the drug's ability to increase red blood cell production and thus the oxygen-carrying capacity of blood that made it an obvious choice for a host of endurance athletes. Historically, athletes have experimented with depressants such as alcohol, stimulants such as caffeine, analgesics such as aspirin, and even more questionable substances such as strychnine to boost their performances, so experimenting with a blood-altering drug such as EPO just extended the tradition.[15]

Athletes were able to procure the substance as early as 1987, but the historical record and the associated confessions across a host of sports have shown that the 1990s became the golden era of EPO use in sport.[16] The rise in the use of blood-boosting drugs during the 1980s and 1990s coincided with the entrance of new technoscientific training techniques into elite sport. The coming together of doping with new research methods on performance mirrored the work performed on anabolic steroids and strength in prior decades.[17] Both moments moved away from the private and nonscientific trial-and-error processes to laboratory-based methods for determining efficacy. Athletes who physically and mentally pushed their bodies to their extreme limits would potentially do the same with performance-enhancing substances, but by these periods they could do so with more technoscientific oversight. Athletes were not the only ones interested in deploying science, technology, and medicine to increase the performance capabilities of the human body.

For example, Francesco Conconi, professor at the University of Ferrara and the Centro Studi Biomedici Applicati allo Sport, and his pupils Michele Ferrari and Ilaro Casoni aimed to replace unfocused experimentation by enterprising athletes with rigorous research. Conconi saw the value of bringing technoscientific precision to sport performance. He wrote, "The modern athlete's adventure is similar to the research scientist's. They both continue to try new things, even at risk of failure, spurred on by the desire to get new results."[18] It was the reconceptualization and use of the human body as a technoscientific laboratory at the Centro Studi Biomedici Applicati allo Sport that also led the world to studying the impacts of EPO on athletes.[19] These researchers laid the foundation to definitively prove that EPO provided game-changing performance gains for athletes in endurance sports.[20] Unquestionably, this group of researchers ushered in a new era of athletic training by elevating it to a verifiable and reproducible science; yet they also blurred the boundaries between technoscientific experimentation and the administration of banned substances.

Ferrari became an outspoken supporter of performance-enhancing substances. He publicly defended his stance by commenting: "If it doesn't show up in the drug controls, then it's not doping." In regard to EPO, he stated: "EPO is not dangerous, it's the abuse that is."[21] His direct and somewhat cavalier attitude reflected the changing tenor toward performance-enhancing substances in technoscientifically mediated sporting communities. In effect, he said that it is only doping if you use enough of a substance to get caught by testing. This is a generous reading of his statement. A much less benevolent interpretation understands Ferrari to say that it is only doping if you get caught. This perspective would lead Ferrari to become one of the most sought-after trainers in endurance sport during the 1990s. It is also partially why Lance Armstrong chose to work with him and defend their relationship even after Ferrari received a lifetime ban from the Federazione Ciclistica Italiana, Italy's cycling federation.[22] Arguably, the lack of restraint by scientists trained by Conconi, linked to the pressure to win, motivated some hopeful athletes to go to the extreme. For athletes, the upside of EPO, besides being undetectable, was the career-altering performance improvement. But the downside of increasing the number of red blood cells is that the blood thickens and blood viscosity decreases. As a result, a high percentage of red blood cells can cause clotting and, eventually, heart failure.

From 1987 to 1989, nearly twenty deaths of young cyclists have been attributed to the experimental use of EPO.[23] The idea that young and extremely fit cyclists who were about to reach their peak years of performance were dying of heart attacks was cause for significant alarm, but this reality did not seem to alter the culture of usage. In sporting cultures demanding extreme physical exertion, the step up from caffeine, to amphetamines, and then to EPO did not seem that radical. It was a calculated risk. In sporting worlds where making calculated risks on the playing field was often a requirement for victory, experimenting with the next unknown performance-enhancing drug could be an immensely productive and lucrative gamble. Part of the commentaries regarding the deaths of the young cyclists focused on how their desire to become champions overrode their common sense and impelled them to take more performance-enhancing substances than their bodies could manage. In many ways, these deaths supported Ferrari's and Conconi's contentions that athletes should have used these substances as part of a training regimen under the guidance of knowledgeable medical scientists.

Most sport governing bodies were cognizant of EPO's existence very early on and quickly moved to ban the chemical. The IOC Medical Commission deemed it potent enough that they placed it on the list of banned substances in 1990.[24] Nevertheless, it would take nearly the entire decade to develop an effective test for this class of drug. The first effective test for EPO appeared late in the last decade of the twentieth century, but the sport of cycling was the first to take action against blood manipulation in 1997. The UCI implemented a new rule governing the upper limit of an athlete's hematocrit level—a measure of the percentage of whole blood volume made up of red blood cells—to 50 percent. Thus if no test for this new blood-boosting drug existed, the UCI decided that it would at least bring the required level down low enough to hopefully prevent substantive enhancement while simultaneously leaving room for athletes with naturally higher hematocrits to compete.

Hematocrit levels vary in all individuals depending on gender, geography, and physiological makeup, but for most humans a "normal" level is below 45 percent.[25] Prior to the legislative change, conservative teams limited their athletes to 54 percent because hematocrits above that mark increase the risk of troublingly high blood pressure and internal blood clotting, though it is rumored that many teams pushed their athletes' physiological limits and went much higher.[26] This physiological cutoff, without the traditional backstop of a verifiable testing procedure for blood manipulation, was an attempt to level the playing field by making sure that all athletes processed similar levels of oxygen.

For cycling, the initial sanction for transgressing the 50 percent mark was light. Riders whose hematocrits rose above the 50 percent level only received the benign reprimand of not being able to start an event until their hematocrit level descended below 50 percent. And the UCI was not the only organization marshaling its resources to fight the hematocrit battle with a "no-start rule." The International Biathlon Union, the International Skating Union, and the Fédération Internationale de Ski experienced substantive increases in athlete performance, which they largely attributed to EPO usage. All instituted rules similar to the UCI's no-start rule to combat the threat of recombinant human erythropoietin in the mid- to late 1990s.[27]

Though the UCI attempted to control its doping problem by requiring riders to slightly change their behavior with hematocrit limits, the 1998 Tour de France Festina cycling team scandal—which exposed the existence of organized doping programs within cycling—made addressing the use of performance-enhancing substances central to cycling's survival. This Tour de France colloquially became known as the Tour de Dopage within the sporting press.[28] On July 8, 1998, in the northern French town of Neuville-en-Ferrain, French customs officers arrested Willy Voet, a Belgian soigneur for the Festina team specifically charged with looking after the team's star rider Richard Virenque, when they uncovered a host of doping products including narcotics, EPO, growth hormones, testosterone, and amphetamine.[29] This stop led to the search and seizure of similar products at the Festina offices in Lyon. In the end, the entire team withdrew from the Tour de France in disgrace, the team disintegrated, and all the members of the team eventually admitted to using banned substances by December 2000. The Festina affair, as well as similar accusations and confirmations of doping within the TVM team, also competing in the 1998 Tour de France, placed an uncomfortable spotlight on doping in cycling.[30] These events exposed the breadth and depth of the doping infrastructure within the sport to international viewing publics.

The Festina affair initially demanded that publics look closer at the sport they loved, become less naive about the realities of doping within sport, and acknowledge that testing is not nearly as effective as sport governing institutions would have us believe. These events also absorbed the public's collective anxiety, disappointment, and anger about doping in sport and allowed other sporting communities to avoid messy public revelations about the sordid realities of their own organized drug and doping cultures. The global profile of the Tour de France, in conjunction with international concerns about performance-enhancing substances, made this Tour de France and, subsequently, the sport of cycling the poster child for all that is wrong with sport and illegal performance enhancement. Cycling became labeled as a sport that accepted and supported the use of performance-enhancing substances. From that moment onward, sporting cultures, communities, and governing bodies disavowed and hid their own issues with performance enhancement because they saw the collateral damage sustained by the sport of cycling. The impact of these events still reverberates in France.

On July 24, 2013, the French senate, in its continued efforts to address the doping problems within their national sport, released a report revealing the names of athletes from the 1998 Tour de France whose samples were retrospectively tested in 2004 for traces of EPO.[31] The idea that the French senate would release information about the Festina debacle fifteen years after the fact clearly indicated the vexing nature of the tension among sport, doping, testing, and national identity. In an effort to wrest the Tour de France away from the specter of doping scandals and reclaim a cherished national event, the French senate opened old wounds and again pushed for a renewed interest in developing tests and procedures that would subdue all the Tour de France's pejorative doping associations. Unfortunately, the report only dragged a host of familiar suspects through the mud again and once more raised the question of if there will ever exist a social, cultural, or technoscientific tool powerful enough to weaken the bond that links cycling and doping.

In other sports, the managing of hematocrit levels appeared to assuage some concerns, or at the very least keep the questions of doping away from nonenthusiast viewers. That all changed, though, during the 2006 Turin Winter Olympic Games, when the Italian police performed nighttime searches of the Austrian cross-country ski and biathlon teams' housing. The IOC and the WADA triggered the searches in the villages of Pragelato and Sansicario because they believed that former Nordic coach Walter Mayer was visiting the Austrian athletes. Mayer's appearance in the Austrian camp caused concern because he had administered blood transfusions to athletes in 2002 and subsequently had been stripped of his coaching responsibilities and banned from the 2006 Winter Games.[32] This all transpired a few days after the suspensions of twelve skiers because of overly high hematocrit levels. It is not a surprise that cross-country skiing, a sport with similar aerobic demands to cycling, has had more than its fair share of performance-enhancing scandals.[33] What makes the appearance of Mayer important is that it illustrates that the new efforts to detect EPO were having somewhat of an effect.

Due to the concerns about testing's ability to detect EPO by 2004, athletes moved away from testable substances to the older, but potentially equally effective, technique of "blood-boosting," or infusing one's own (autologous) or another's (homologous) red blood cells to increase oxygen-carrying capacity. The process had fallen out of favor with the development of EPO and the simplicity of administering the drug compared to the infrastructure required for blood boosting. With EPO, athletes could manage the purchasing, storing, and injecting on their own, whereas it was nearly impossible to blood boost alone. The basic process of blood boosting requires an athlete to withdraw a quantity of blood at least six weeks before a targeted competition. Then, the athlete, or another knowledgeable individual, separates the red blood cells and stores them in a refrigerated location. In the subsequent weeks, the body replaces the extracted red blood cells, and shortly before competition the stored red blood cells can be injected into the blood stream to increase the athlete's red cell count and oxygen-carrying capacity. The process requires someone with phlebotomy skills, an understanding of red blood cell harvesting techniques, and a secluded storage facility for the cells. Needless to say, this approach demands a level of technoscientific infrastructural support.

It is believed that five-time Tour de France winner Jacques Anquetil (1957, 1961–1964) was one of the first athletes to experiment with this technique.[34] Anquetil, who dominated cycling in the early 1960s, was quite outspoken about his drug use and how it was a necessary tool if fans wanted to see amazing cycling and athletic feats.[35] Anquetil did not mince words when he stated that "you'd have to be an imbecile or a hypocrite to imagine that a professional cyclist who rides 235 days a year can hold himself together without stimulants."[36]

Blood boosting rose in interest among athletes after the 1968 Mexico City Olympic Games. Scientists showed that endurance athletes who trained at altitude prior to the 1968 games performed significantly better than athletes training at lower altitudes.[37] The value of having more red blood cells began to infiltrate multiple sporting cultures. The efficacy of blood infusions was well established by the 1970s, when a group of researchers led by Björn T. Ekblom validated that "after reinfusion [of blood] there was an 'overnight' increase in . . . physical performance capacity."[38] This seemingly benign statement scientifically confirmed sporting's common knowledge that more red blood cells produced better performance. Future research even dismissed any health concerns about the process by concluding that "there has been no report on any complication in connection with reinfusion using autologous whole blood or packed RBC [red blood cells]."[39] Moreover, it did not appear to many to be problematic when an athlete used his or her own blood. It was not as if an athlete injected a foreign substance or product. The athlete's body had produced the red blood cells.

Blood boosting became a dominant form of performance enhancement because it was relatively simple if one had the infrastructural support, it had not been proven to have adverse health effects, it was difficult to determine if an athlete had used the procedure, and it resided in an indeterminate ethical space. Again, many athletes did not see this procedure as doping because it did not require them to use some foreign or illegal substance produced in a laboratory. They were just using blood, a natural product created within the human body. Whether it was their own or another's, however, the introduction of more red blood cells would eventually carry the same pejorative weight as banned performance-enhancing substances.

In 1983, Kenneth Clarke, then the United States Olympic Committee medical director, attested to the proliferation of the technique when he stated that blood boosting "is fairly widely practiced in Europe, especially among cyclists and Nordic skiers, [but] . . . the use of someone else's blood is now clearly verboten." Some may have found it to be unethical, but in regard to using one's own blood Clarke stated: "From a medical point of view, it can now be considered ethical. However, no organization, the IOC especially, has ever clarified the ethical value of IE [induced erythrocythemia, or the introduction of a surplus of red blood cells] in sport."[40]

The procedure mostly remained out of public sight until the 1984 Los Angeles Olympic Games, where seven of the twenty-four-member United States cycling team participated in a blood-boosting plan suggested by national team coach Eddie Borysewicz and under the guidance of Herman Falsetti, a professor of cardiology at the University of Iowa. This practice probably would have remained in the shadows if Mark Whitehead had not fallen ill, potentially due to a blood mix-up. Whitehead's own blood was unavailable for the Olympics, so he chose to have a homologous blood transfusion. This choice resulted in a fifty-hour fever of 103 degrees and, subsequently, the unraveling of his Olympic dream. The IOC reacted quickly to the negative publicity and banned the procedure in 1985, even though it did not have a definitive test to determine if an athlete had blood boosted.[41] In 1999, after decades of public performance-enhancement debacles that questioned the IOC's ability to govern over and facilitate clean sporting competitions, the IOC created the WADA. The WADA became a partner institution designed to curtail the dwindling belief in the IOC's power to police doping in Olympic sporting competitions as well as a method to outsource the responsibility of monitoring, testing, and sanctioning doping infractions.

The Rise of the World Anti-Doping Agency and the Athlete Biological Passport

The idea of an international doping organization had been percolating for some time in the late twentieth century and finally boiled over as the millennium approached. The time was right for an overarching organization to manage and coordinate the international network of anti-doping efforts. Each nation and their various sport governing bodies had their own rules and regulations regarding banned substances, testing procedures, and doping sanctions. Though collaboration was desirable, cross-organization cooperation was erratic at best, which made it challenging to have consistent international policies on things such as lists of banned substances, testing protocols, and punishments for offenses. In February 1999, the IOC took the lead and hosted the first World Conference on Doping in Sport in Lausanne, Switzerland. The main result of the meeting was the adoption of the Lausanne Declaration on Doping in Sport.[42] The assembled international sport federations, national Olympic committees, and athletes agreed to a statement making six declarations.

The first declaration, "education, prevention and athletes' rights," pushed for a better system of informing athletes about the dangers of using performance-enhancing substances. The idea behind this declaration was that if governing bodies could educate athletes early in their careers, or even before they seriously committed to athletics, they could motivate a cultural shift away from doping by having young athletes embrace a given sport's history and tradition of ethical, fair play.

The second declaration, "Olympic movement anti-doping code," required that all athletes and governing institutions abide by a uniform code and list of banned substances. This was an important step because it universalized the expectations of all athletes regarding doping, the appropriate responses by sport governing bodies, and the all-important list of banned substances.

By agreeing to the code, the athletes and governing federations also accepted the third declaration of "sanctions." The group agreed that the minimum suspension for a major doping infraction would be two years. Agreeing on identical doping penalties was a major step forward and eliminated uncomfortable comparisons between sports, nations, and governing institutions.

The fourth declaration was the most profound because it called for the creation of an "international anti-doping agency" to be in place by the 2000 Sydney Summer Olympic Games. The IOC committed the first $25 million to this endeavor. This last action item materialized on November 10, 1999, with the creation of the WADA.

The WADA would possess much power, but the fifth declaration, "responsibilities of the IOC, the IFs (International Federations), the NOCs (National Olympic Committees), and the CAS (Court of Arbitration for Sport)," also bolstered the governing authority of these national organizations. This fifth declaration maintained that each country's existing institution or organization charged with evaluating doping cases would keep this responsibility. Thus each country's anti-doping organization would be responsible for policing its own athletes. Once a national organization sanctioned an athlete, his or her last option would be an appeal to the CAS, which would render the final judgment.

Finally, the sixth declaration, "collaboration between the Olympic Movement and public authorities," proclaimed that anti-doping was a collective effort, and institutions as well as the public must participate in efforts to curtail doping in sport. The Lausanne Declaration on Doping in Sport authorized the WADA with the charge of organizing and supporting a broad-based anti-doping network. In its effort to refresh the anti-doping movement in light of revelations regarding the limitations of direct testing, the WADA began discussions about creating a new anti-doping device, instrument, or system of measurement as early as 2002. Mario Cazzola, a professor at the University of Pavia School of Medicine, was one of the first to propose a "hematologic passport" in 2000.[43] As a member of the Federazione Italiana Giuoco Calcio's (Italy's soccer federation) Science Committee, he created the "Io non rischio la salute! [I take care of my health!]" program to monitor the blood values of Italian soccer players. A group of researchers led by Robin Parisotto of the Australian Institute of Sport published some of the early research confirming the potential of indirect testing for blood manipulation in 2000 as well.[44] But it was not until 2006 that the science, technology, and cultural impetus all coalesced to lead to a solution: the ABP.[45]

In 2006, the WADA convened the Haematological Working Group to begin mapping out a plan to create an "Athlete's Haematological Passport." Pierre-Edouard Sottas was one of the leading research scientists tracking abnormal blood values.[46] Some of Sottas's published work built the foundation for the ABP. Sottas and his team made a case for "a new indirect test, called Abnormal Blood Profile Score, that is sensitive to multiple blood doping practices."[47] Their studies showed that it was possible to use statistical classification techniques to develop a more comprehensive method of determining if an athlete used blood-boosting drugs such as EPO or increased their performances with blood transfusions. This was a monumental step forward because historically it had been extremely difficult to catch blood manipulation. Sottas, who also served as the manager of the ABP for the WADA, and his co-authors saw the ABP as a means to reclaim the integrity of sport from the clutches of doping. Early on, there was great excitement about the potential of the ABP, through its ability to longitudinally track blood values to severely curtail all forms of doping that involved increasing the oxygen-carrying capacity of blood.

In October 2007, the UCI and the WADA agreed to collaborate and launch the ABP in cycling for the 2008 season. As a result, the 2008 racing season began with the great hope that the UCI would be able to guide cycling out of its recent dark past with the implementation of the ABP. When initially introduced to the ProTour cycling teams (the eighteen highest-ranked teams in the world) and a few of the ProContinental teams (the second tier of professional cycling, which strived to garner the "wildcard" slots at elite races such as the Tour de France), the athletes begrudgingly complied. The teams contended that the ABP was an unwelcome and unsanctioned invasion of privacy. Nevertheless, the UCI and the WADA proceeded with a plan to collect ten blood samples and four urine samples, during events and out of competition, from each elite professional cyclist under their jurisdiction. The testing tracked the following parameters: hematocrit (HCT), hemoglobin (HGB), red blood cell count (RBC), the percentage of reticulocyte (RET%), reticulocytes count (RET#), mean corpuscular volume (MCV), mean corpuscular hemoglobin (MCH), and mean corpuscular hemoglobin concentration (MCHC). By most accounts, the process succeeded.

At the end of the 2009 season, the UCI and the WADA had amassed complete samples from 804 of the 848 athletes competing.[48] The acquisition of this dataset was critical for building a case for the efficacy of an ABP system. These cyclists represented a set of the highest level of endurance athletes. Thus the data gave researchers, the UCI, and the WADA a comprehensive snapshot of the blood values of elite endurance athletes, and certain causative connections about the range of normalized blood values could be deduced from this data. This process of collecting biological data also let athletes know that their blood values would be tracked and compared to themselves and potentially their athletic compatriots.

This new type of testing supported by historical corroborative data also potentially meant that prior techniques for evading sanction would substantially disappear. Every blood test would be compared to an athlete's own historical blood values, and, ironically, if an athlete's blood values were deemed to be abnormal and worthy of sanction, his or her own body would be significantly responsible for revealing the blood manipulation. Thus the ABP turned athletes' own bodies into disciplining and policing instruments. The UCI and the WADA were optimistic that the ABP would send a strong signal that they were redoubling their efforts to reform sport. At the beginning of 2008, a strong sentiment existed that the ABP would change athletes' behavior and give the public a reason to believe that the sport of cycling, and by extension all sports, had put into place an effective deterrent system for illegal blood manipulative substances. The prevailing attitude was that the ABP would swiftly close the gap between those using the next potent performance-enhancing substances and those trying to end their use within sport. But by the end of the summer, this prediction, though partially correct, showed how far the ABP needed to go if it wanted to become the magic bullet for cycling and all endurance sports.

The Athlete Biological Passport Comes of Age

When the 2008 Tour de France ended on the Avenue des Champs-Élysées in Paris in late July, most would agree that the ABP had made an impact. The media fallout from the removal of stage winner Riccardo Riccò from the race (he was caught using CERA, or continuous erythropoietin receptor activator, the latest EPO-based, blood-boosting drug) and the withdrawal of his entire Saunier Duval team clearly made this point.[49] The fact that athletes were using CERA also highlighted the difficulty that the ABP would have in managing the science and politics of the ever-changing world of performance-enhancing drugs. Roche Pharmaceuticals began marketing their CERA drug (epoetin beta and methoxy polyethylene glycol injection) for the treatment of anemia associated with chronic kidney disease under the trade name Mircera early in 2008. The primary chemical difference between CERA and EPO is the addition of polyethylene glycol. The drug received approval from the European Commission and the United States Food and Drug Administration in 2007.[50]

In clinical trials, this longer-lasting form of the famed erythropoietic-stimulating agent had proven to produce a more stable increase in red blood cells. Clearly, a longer-lasting form of EPO was highly attractive to endurance athletes. Rumors circulated that the increased molecular mass would make the drug undetectable in urine. Yet as positive tests from Riccardo Riccò, Leonardo Piepoli, and other cyclists indicated, this hope was far from accurate. Even though the ABP exposed athletes that should receive additional scrutiny and eventual testing, catching these athletes required the cooperation from an international pharmaceutical company.[51] L'Agence Française de Lutte contre le Dopage, the French doping agency, succeeded in catching these transgressions by selectively targeting athletes with questionable blood values during prerace screening, but more specifically through the use of a new direct blood test for Mircera developed in conjunction with Roche Pharmaceuticals.[52]

Inspired by this cooperative relationship, the International Federation of Pharmaceutical Manufacturers and Associations (IFPMA) and the WADA formally agreed to join forces on July 6, 2010, by signing the *IFPMA-WADA Joint Declaration on Cooperation in the Fight against Doping in Sport*. This agreement encouraged the sharing of information and resources in support of anti-doping efforts. The remarks at the announcement by WADA director general David Howman clearly indicate that he firmly knew what was at stake and saw it as a significant achievement for the anti-doping movement. In this regard, he stated:

> Cooperation with the pharmaceutical industry is a win-win for WADA and the pharmaceutical industry. It helps the anti-doping community further tighten the net on doping cheats. At the same time, it helps the pharmaceutical industry ensure that their work is directed at treating and healing patients suffering from illness and disease, not at providing healthy athletes with an unfair advantage over their competitors. Doping is a public health issue, and WADA and the IFPMA share the objective of promoting and protecting public health.
>
> Much of the doping that occurs in sport is the misuse by cheating athletes of medicines that were or are being developed by the pharmaceutical industry for proper medical use. There are thousands of new drugs that are at various stages of pre-clinical and clinical development. These new drugs will hit the market in the years to come and

some might be used for doping purposes. It is therefore very important for the anti-doping community and the pharmaceutical industry to be in a position to address this issue together, in a coordinated and effective way.[53]

Haruo Naito, then-president of the IFPMA, mirrored Howman's remarks and inveighed: "It is our hope that this declaration will give rise to a growing range of bilateral agreements [that] should slam the door shut in the face of dopers who hope to abuse individual candidate medicines which are still in development." He strongly asserted that this institutionalized cooperation was necessary "to avoid giving early warning to dopers," and that it was equally important "to be rather discreet about these individual agreements, at least in their early stages."[54]

Though voluntary, it was good business for all. The pharmaceutical industry was coming under fire because it, intentionally or not, was the source of performance-enhancing substances. Most companies created these substances as valuable technoscientific treatments for a host of medical ailments and had little control over how athletes and trainers transmogrified these substances to enhance athletic performance. Pharmaceutical companies were as keen on curtailing the illegal use of their substances as the WADA, as its connection to doping was bad for business. This collaboration gained strength on June 28, 2011, when the global Biotechnology Innovation Organization (BIO)—the largest trade organization representing the biotechnology industry—endorsed the *IFPMA-WADA Joint Declaration on Cooperation in the Fight against Doping in Sport*.[55] On July 23, 2012, the IFPMA, the BIO, and the WADA took another organizational step and launched the "2 FIELDS 1 GOAL: Protecting the Integrity of Science and Sport" campaign to build "a strong framework of collaboration and encourage the voluntary cooperation of IFPMA and BIO member companies with WADA to readily identify compounds with the potential for misuse by athletes and to stop doping in sport."[56]

Though the IFPMA, the BIO, and the WADA agreed to share information about emerging performance-enhancing substances through the 2 FIELDS 1 GOAL campaign, there may always be more resources to find ways to administer and hide the use of these substances than there will be to detect them.[57] For erythropoiesis-stimulating agents such as Mircera, the likelihood of similar industry-wide cooperative actions will be difficult because not only is direct testing the solution and the problem, but these cooperative testing protocols need to address the fact that "over the past 2 decades . . . 6 different rEPOs have been licensed worldwide and more than 90 biosimilar rEPOs or copies have become available in countries with low regulatory controls of pharmaceutical products."[58] The proliferation of these and similar products makes the detection of all substances untenable. Part of the problem is that new and more effective performance-enhancing substances are continually being developed. In a sporting world where direct analytical detection is the primary way of determining an infraction, savvy athletes will always attempt to find and use substances that do not appear on the WADA's banned list.

Direct detection, as a method, always will have an intimate connection to the banned substances list. But unfortunately for the anti-doping network, the list probably will always be incomplete. The history of sport is dotted with instances in which tests indicated that an athlete used a substance that can increase performance but that the product was not officially on a banned substance list. The UCI indicated that the ABP had been effective in changing the behavior of cyclists, and their blood value data proved this, yet this perception could also be reread as a narrative of the ways astute athletes move on to greener doping pastures. Even for athletes with less sagacious dispositions, the inability of the ABP to effectively detect microdosing of EPO, autologous transfusions, and a host of other forms of blood manipulation does not give many scientists a reason for great optimism.[59] Consequently, there are still ways to avoid the controls set forth by the ABP system.

Naming this instrument a passport conjures a host of images about cosmopolitan travel, national identity, and transportation gateways. It taps into traditional representations of necessary stamps of approval from a governmental body. The UCI and the WADA designed the ABP as an instrument to govern which bodies would have access to the sport of cycling, and eventually all sports. The ABP, as a technoscientific instrument, became a proxy to enforce the UCI's control over riders and to protect its financial interests from the negative effects of drug use. Anne Gripper, the former anti-doping manager of the UCI, described the ABP as "a series of tests that enable us to make a determination as to the likelihood of doping based on that rider's individual profile. So rather than comparing one single sample to a population norm, we're comparing a range of samples to an athlete's expected profile. So it gives us a lot greater sensitivity, enabling us to determine that this rider is likely to be doing something that manipulates their blood, or likely to be doing something that relates to steroid use."[60] By acknowledging that the ABP creates a specific profile of blood values that can assist in determining guilt or innocence, Gripper affirms that the UCI "may not actually be able to say what it is, whether it's autologous blood transfusions or micro-dosing with EPO, but what it will show is that this rider is highly likely to have been doing something illegal. So it's a whole new approach; it's using that forensic approach, assessing evidence to the point where you believe you've got a quality set of data that can take us to the use or attempted use into doping."[61]

Herein lies one significant issue for the ABP that it may not be able to technoscientifically resolve. An individual athlete's longitudinal blood values can *allude* to wrongdoing—which is significantly different than being caught red-handed with an illegal substance within one's body. Those institutions deploying the ABP are comfortable with the fuzzy level of statistical certainty the passport provides. But as the ABP system is rolled out across multiple sports, will probabilistic certainty convince the public that their beloved sports are any cleaner? The technoscientific, social, and cultural use of the ABP connects to the development of forensic science. In speaking about the ABP, Gripper carefully used the term *forensic science*, and Sottas's writings have promoted forensic science as a key element for the new anti-doping movement. The ABP approach of characterizing anti-doping testing and enforcement as forensic science, for all intents and purposes, gives up on direct testing. Thus it appears that the commentaries by Gripper, Sottas, the UCI, and the WADA constructed a reformation narrative that allowed for the failure of the direct testing model while simultaneously presenting a seemingly more effective tool in the ABP.

When the WADA distributed the *Athlete Biological Passport Operating Guidelines* in January 2010, it contained explicit details about sample collection, transportation, and storage. This document articulated a rigorous sample management protocol to assuage concerns that this new indirect method was fully trustworthy. The WADA designed this procedure to meet a forensic science standard. This connection of the ABP to forensic science is critically necessary because the ABP can be easily interpreted as less conclusive than direct testing. It moved away from the hard direct testing data to the more nebulous process by which "evaluation consists of a longitudinal assessment of biological variables to determine the probability of the data being physiologically on the basis of the athlete's own previous values (performed by an automated software system using a Bayesian model) and a subjective evaluation of the results in view of possible cause (performed by experts)."[62] This process is a long way from the perceived decisiveness associated with positive analytic tests. The forensic science language is invoked to refresh the ABP process with new technoscientific garb. The hematologist for an ABP case is analogous to a forensic scientist in a criminal case. Both experts review carefully constructed evidence to produce an evaluative recommendation. The WADA strongly emphasized and described the quality control at WADA-certified laboratories as being above the clinical level and at the forensic level.

The problem is, how does an organization give up on direct testing without admitting decades of failure? The ABP provides an out, but not necessarily in the eyes of the public. This is why athletes such as Ben Johnson, Lance Armstrong, and Barry Bonds can be so polarizing. The public wants to know the explicit details of performance-enhancing drug use to make up their own minds regarding guilt or innocence. Direct testing has been useful in this regard, because it has been effective in allowing all sides to argue a position. If an athlete, no matter how high the suspicions are regarding performance-enhancing drug use, does not test positive, the athlete, their fans, and their sport can claim that all is well. If the athlete does test positive, their competitors, their former fans, and their sport can unceremoniously toss them into sport's abyss of dispossessed athletes. Since the anti-doping movement traditionally has cast its lot with direct testing for a host of relevant reasons, the ABP—a seemingly softer form of testing—does not seem to be a step in the right direction of providing more convincing evidence. Though the ABP improves the processes by which governing bodies surveil athletic bodies, the complexity of the testing regime may make it difficult to convince sporting publics, who traditionally like clear decisions (e.g., the ball did or did not go through the goal and we all saw it), that this instrument will do a better job of cleaning up sport and routing out those who undermine its traditions by cheating.

Part of the strength, and weakness, of the ABP is that expert analysis can be informed by disciplinary expertise, or undermined by jaded disciplinary ignorance. Just as the science seems to get murkier, the

ABP moves the enforcement location of clean athletic performance deeper into technoscientific laboratories. The processes by which clean performance is determined are getting increasingly black-boxed by the technological sophistication of the diagnostic tools and instruments and the breadth of scientific knowledge required to interpret these samples, observations, and data. The UCI and the WADA, and other sport governing organizations implementing the ABP, consider this to be a good thing. They have called this new form of technoscientific monitoring a turning point. The UCI has emphasized that new technoscientific rigor can effectively reveal offending behaviors, yet its confidence is couched within an amorphous technoscientific trust. It contends that the ABP will succeed because "it draws upon important new scientific methods of indirect detection, . . . uses sophisticated statistical tools to interpret results, . . . [and] uses a sequence of tests to provide greater sensitivity in testing."[63] Even though the UCI has had issues with the transportation, handling, and storing of athlete samples, it contends that these difficulties do not undermine the ABP program in part or as a whole.[64]

Moving forward, the WADA and the UCI have committed to non-analytic testing procedures as a fundamental part of the tests administered to athletes. The WADA has plans to implement a suite of nonanalytic testing protocols and is developing a steroidal module (for direct and indirect detection of anabolic agents) and an endocrinological module (to detect human growth factors), but the hematological module of the ABP is currently the only module in use.[65] The WADA sees the development of these tools as the essential building blocks of a comprehensive anti-doping regime. For the WADA, these steps are necessary because it is becoming increasingly difficult to determine if an athlete has used a banned substance at a specific moment in time, which means that these cases can be difficult to prove because indirect evidence can be viewed as circumstantial.[66] An athlete can have a nonanalytical positive for a large variety of reasons, but will publics support the banning of an athlete even when a direct positive test result does not exist?[67]

This new testing world calls for a cognitive metamorphosis where athletes, governing bodies, and publics all agree, for the most part, that an athlete can be suspended or permanently banned from competition in an instance in which the traditional marker of a violation—a positive test—

is not present. At the present moment, the WADA may be asking too much from this group of constituents. More significantly, recent studies have pointed to fundamental limitations in the ABP and its reliance on Bayesian modeling to determine illicit behavior.[68] Specifically, a group of researchers led by Thomas Christian Bonne at the University of Copenhagen determined that altitude training adversely affects the accuracy of the ABP data. They studied a group of elite swimmers who lived and trained at high altitudes (over 2130 m) for at least three weeks. After the three weeks of training, the researchers found that 60 percent of the athletes "exceeded the 99% ABP thresholds," which, according to the ABP, would mean that they had manipulated their blood.[69] The athletes did alter their bodies' oxygen-carrying capacity, but not with banned substances. They went "old school" and used altitude to garner similar effects, which may have caused them to run afoul of the ABP.

Scholars in the field of science and technology studies have shown that the production of scientific knowledge and technological instruments is a social process embedded within an ever-changing network of cultural interactions.[70] Drawing from this scholarly tradition, the ideas of getting a better, truthful, or more authentic athletic performance by increasing the scientific and technological oversight seems fraught with problems. This scholarship has championed the social construction of scientific knowledge and technological artifacts by exposing the social and cultural nature of scientific and technological work. Unfortunately, this can often be the same approach used by those accused of committing various offenses. For instance, an athlete's first response to a positive drug test is to claim that someone made a mistake or that the test recorded inaccurate data. The goal of showing the contingent nature of science and technology may have produced the idea that reality does not exist and truth is always negotiable. This is an important scholarly project, but it has troubling ramifications for sport. Thus how do we find a middle ground to understand, evaluate, and critique athletic performances mediated by emerging technoscience and its newly developed diagnostic instruments? It appears that in the near future, endurance sports such as cycling, cross-country skiing, and distance running will be as much about tests of human endurance and performance as they will be about testing the efficacy of scientific knowledge, diagnostic protocols, and technological instruments.

A passport is a gatekeeping device. The proper one facilitates entrance, while the wrong one, or none at all, denies access. The WADA and the UCI developed a new document, the ABP, to reclaim testing from the ashes of its smoldering history of failure. Those supporting this instrument are potentially making a significant mistake. By pushing testing deeper into the hidden spaces of the laboratory and further away from a curious and invested public, the ABP creates a new technoscientific regime that misses an opportunity to add a level of transparency in addressing the timeworn question of who can and cannot compete or who should and should not be sanctioned. Unfortunately, the anti-doping movement seems to think that the technoscience will speak on behalf of the movement, but, sadly for them, the ABP is voiceless. Though relatively new and ingenious, the ABP does not appreciably address the grayness found at the moral, philosophical, and ethical edges of sport.

For instance, the psychological and practical distance between medications for health and recovery, and the substances that improve performance is small and varies in the minds of elite athletes and society at large. The process of depletion and replenishment is a never-ending activity for elite athletes, and the ability to recharge one's body is paramount to continued success. Thus testing unfortunately creates the hard and fast line between legal and illegal. In practice, this clear demarcation does not exist, but athletes are tested and sanctioned as if it is plain for all to see.[71] Of course, governing bodies intend for the list of banned substances to occupy this void, but the list is no more stable than anything else. For instance, the stimulant caffeine was removed from the WADA list of banned substances in 2004.[72] The list is never complete; however, it is often presented as definitive. This is the focus of the tension around testing. Governing bodies and publics live by clearly defined lines of accepted and banned substances, legal and illegal bodies, permissible and impermissible technoscience. But for many athletes and fans alike, these parameters do not coincide with the realities of their sporting lives.

The ABP, as a material artifact, illustrates the tensions this form of technoscientific surveillance produces for athletes, sporting cultures, and the governing bodies legislating its use. The builders of the ABP systems hope this device will induce the public to more fully trust athletic performances in the future. That is, the effectiveness of the ABP matters only as long as governing bodies such as the UCI, FINA, and the WADA can get the public to trust the passport and, subsequently, the athletes and sports they govern. The reliance on an increasingly complicated system to technoscientifically oversee biological data may undermine their fight to create a standard by which athletes can perform a legislated form of athletic authenticity. Thus only time will tell if this technoscientific instrument will resolve this long-standing social and cultural problem.

Body, Motor, Machine

The Future of Technology and Sport

We love champions. We especially love sport champions. We love heroes. We especially love to make athletic success heroic. This love readily conflates heroism and athletic success. The heroic champions we love most are those who have worked the hardest and sacrificed nearly everything to achieve the highest heights of success. Winning a sporting competition always carries a level of status, and athletes can drive themselves to unhealthy and irrational lengths to win and be loved. The apocryphal legend of the Greek messenger Pheidippides's nonstop run from the battlefields of Marathon to Athens in 490 B.C. exemplifies the value in the reward and the associated heroic narrative.[1] As the story goes, Pheidippides collapsed after exclaiming "nenikekamen," or "we have won." This compelling narrative also served the organizers of the first modern Olympic Games in 1896. The Pheidippides legend supported making the marathon, as a death-defying twenty-six-mile run, a centerpiece of the 1896 games. Pierre de Coubertin, the instigator of the modern Olympic Games, clearly understood the promotional value of this event. And his intuition was spot on, as the marathon has become one of the signature competitions of the Olympic Games. What makes the marathon so compelling is its history, the palpable physical struggle, and the perceived purity of running.

This athletic purity has defined many of the most successful sporting competitions as they moved from recreational pastimes to professionalized events after the turn of the twentieth century. This transition can be seen most poignantly in boxing. Boxing also helped turn sport into sports business. Celebrated boxer Jack Johnson is not only known for his athletic accomplishments in a racially adversarial world but also for the associated heroic narrative that came with being a championship-winning athlete.[2] It was the watching and cheering for and against athletes doing the seemingly impossible, and potentially placing a wager on it, that made these athletes and their performances exciting for fans and lucrative for the athletes and the various event promoters. The packaging of these sporting events laid the foundation for one strain of sporting consumption that hinges on marketing and selling great human athletic feats. It is the promotion of athletic ability as a consumable product that links Pheidippides and Jack Johnson to the contemporary multilevel athletic marketplace, where anyone can consume an event, a team, or a player. But in this current moment, is it still possible to uphold the heroic athletic narratives represented by Pheidippides?

The athlete and the athletic body will always be central to this form of sport consumption. There are powerful social and cultural reasons why we speak of sporting competition in terms of human dramas. Sport can be very dramatic; thus, it is no surprise that sporting events often are construed as a battle between good and evil, right and wrong, democracy and communism, and so forth. Rooting for one's "side" can make being a fan worthy of extreme and outlandish behavior. This support can also permit fans to ignore the transgressions of a favorite athlete or team, as well as bestowing one's side with superhuman abilities. It is the idea that some athletes have "it" while others do not that effectively works to marginalize the importance of the technoscientific infrastructure of sport. Though it is clear that an extremely select few athletes, such as Michael Jordan, Lionel Messi, or Wayne Gretzky, possess that rare combination of genetic ability, motivation, and luck to dominate a sporting domain in a specific historical moment, the physiological differences between elite athletes are exceedingly small. But what if great athletes succeed because they more readily adapt and exploit the old, new, and emerging technoscience of their sport?

For more than a century, the dominant narratives around science and technology have been highly progressive. These narratives are not without their critics, but overall the idea that the confluence of science and technology will create a "better" technoscientifically driven world is very strong. Historically, sport has been able to hold on to a humanized narrative and sublimate all things technoscientific to the body. However, as the preceding chapters have shown, technoscience now plays a central role in the outcomes of sporting competitions, and to understand the impacts of technoscience on culturally rooted sporting narratives, the human body, and its meaning for sport is a key starting place. What makes the body so important to culturally valued sporting narratives is the perception of its purity and, most importantly, authenticity of performance.[3]

Concepts of bodily authenticity are deeply meaningful for sport because sporting narratives are built on the belief that bodies decide who wins and who loses. Ultimately, fans cheer for athletes and not machines, artifacts, or devices. In the recent past, sports that have focused their narrative histories on the primacy of bodily athletic ability have struggled to maintain this human-centered vision. Arguably, a considerable part of this struggle has been the collective denial that modern sport, from its earliest instantiations, has been technoscientific to a degree, just as much as the world in which we live is technoscientific. The "tools" and equipment of sport can no longer be understood in instrumental terms as benign artifacts of the game. The long-cherished belief that the only way these artifacts influence competition is through the hands, minds, and bodies of skilled athletes must give way. These technoscientific artifacts are much more than instrumental tools. They are devices that in the recent past have pulled sporting outcomes decidedly away from the competitors, governing institutions, and fans to the technoscientific.

As these sport governing bodies, consuming publics, and athletes work to extract and diminish the technoscientific realities of sport, technoscience has and always will be an integral component of sporting culture. Until the last thirty years of the twentieth century, this nontechnoscientific illusion was easier to maintain. With the fine-tuning of performance-enhancing drugs, developments in material science, and evolution of engineering design, the efficacy of sporting technoscience has become harder and harder to ignore, disavow, and legislate away.

We are, for all intents and purposes, at a precipice where the differences between the bodily abilities of athletes in a sporting domain may become so infinitesimal that athletic performance may cease to determine the outcomes of sport competitions. This ironic instrumentalization of the human body will push athletic competitions further along a pathway to where the most heated competitions no longer take place between humans on the playing field but between humans in scientific or engineering laboratories where a sporting body becomes another medium through which to display the latest scientific knowledge or technological innovation. The examples presented in this book aim to lay a foundation to impel the rethinking of humanity's relationship with technoscience in the context of sport in order to begin a conversation about where the limits of technoscience within sport can and should be drawn. The examples herein also provide a basis for building a framework from which to begin these all-important sporting and cultural dialogues.

Though there are many things that influence sport and the outcomes of sporting competitions, the set of examples discussed emphasize the power of technoscience and the ways in which it impacts perceptions of sporting bodily authenticity maintained by sport governing bodies, invested publics, athletes, and the technoscientific actors who design, build, and sell the next game, as well as redefine artifacts, objects, and equipment. At a most basic level, the tensions around these relationships can be represented graphically (figure 7.1). This schematic does not mean to imply that these relationships are static. Though these interactions continually

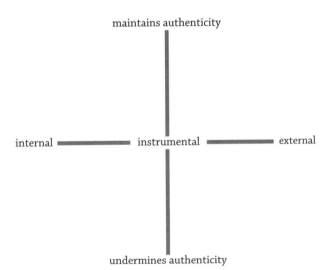

Figure 7.1. The authenticity/body coordinate system illustrates the overlapping ways different technoscience supports and undermines forms of sporting and cultural authenticity.

morph into new arrangements, this diagram simply illustrates how perceptions of technoscience migrate and subsequently influence the ways in which different sporting cultures conceptualize bodily authenticity in relation to technoscience. This coordinate system can also be used as a way to locate and mark the sporting cultural stakes for existing, new, and emerging sporting technoscience.

Internal, Undermines Authenticity

Beginning in the bottom-left corner is technoscience that is hard for the naked eye to see, but it also undermines an agreed-upon form of sporting bodily authenticity. Technoscientific interventions residing in this space are internal to a body and alter a body's physical ability. The best example of such technoscientific products fall under the category of performance-enhancing drugs and "doping."[4] The full-scale laboratory research into synthesizing these types of compounds began in full after World War II. Initially designed to create stronger and more durable military fighters, trainers and athletes quickly realized and exploited performance-enhancing drugs' effectiveness in sporting competitions.[5] Nearly all sport governing bodies now ban the most potent of these chemicals. Performance-enhancing drugs are not only powerful because of the physical and psychological boosts they give athletes but also because the naked eye generally cannot detect their use. The sheer invisibility of performance-enhancing drugs is part of what makes them so effective for competitors and so troubling for fans and sport governing bodies. One cannot visually discern whether an athlete has consumed, inhaled, or injected an illegal level of a performance-enhancing drug.

This is part of the magic, and problem, of performance-enhancing drug use for sport governing bodies, athletes, and fans alike. Each of these groups can act as if only bad or immoral athletes use performance-enhancing drugs, if they so choose, but as the sporting news of the early twenty-first century has shown, performance-enhancing drugs have been part and parcel of most sports for quite some time, and modern sport and performance-enhancing drugs have developed in lockstep. As troubling as performance-enhancing drugs are for maintaining the belief that the human body is vastly more important than sport technoscience, or the motor-over-machine parable, performance-enhancing drugs also help sport become a network of global, multi-billion-dollar industries by providing greater and greater athletic performances. The public invisibility of performance-enhancing drugs allows fans to simultaneously celebrate the greatness of human ability and maintain the illusion that unadulterated physical ability governs sport.

Sadly, sport governing bodies also directly and indirectly support this mythology. Since sport governing bodies do not and probably will never have the staffing or financial resources to combat those using and profiting from performance-enhancing drugs, the goal is to manage the problem, which historically has meant selectively finding and exposing the performance-enhancing drug problem within a sport. Governing bodies that aggressively police the level of drug use in their sport can pay a high price if the sport becomes known as having a drug problem, as in the case of professional cycling. Since beginning its late-twentieth-century movement to curtail drug use in cycling, the Union Cycliste Internationale contends that cycling has been an extremely clean sport and that the only reason drug testing catches so many cyclists is because the UCI tests more than any other sport.

This line of argumentation may be true (though the Lance Armstrong saga raises significant questions about the veracity of the UCI's claims about maintaining a clean sport), but only serious cycling fans following the sport closely enough can appreciate the in-group technical distinctions between the vigilant monitoring of doping and a sport with a drug problem. Most sport governing bodies prefer to keep testing out of the spotlight and only let performance-enhancing drugs rise to the level of a problem when an athlete's positive test becomes a media event. This is also a moment when sporting cultures avoid acknowledging the functional reward structures and the historical motivations to use substances that enhance performance. In an effort to distance a sport from a culturally abhorrent practice like cheating with drugs, governing bodies and fans castigate drug-positive individuals as aberrations in an otherwise pure sport. This approach can be seen in the ways in which Major League Baseball packaged the 2005 "Restoring Faith in America's Pastime: Evaluating Major League Baseball's Efforts to Eradicate Steroid Use" and the 2008 "The Mitchell Report: The Use of Steroids in Major League Baseball" hearings before the United States Congressional Committee on Oversight and

Government Reform, as well as suspensions of celebrated players such as Alex Rodriguez in 2014.[6]

The place of performance-enhancing drugs within the authenticity–body coordinate system has not always been the same. For instance, the evolving group of drugs that fall under the term *amphetamines* trickled onto lists of banned substances well past the mid-twentieth century, whereas the World Anti-Doping Agency removed caffeine from its list of banned substances in 2004.[7] Thus the meaning and understanding of what is and is not a performance-enhancing substance can change over time. In the case of caffeine, some still call for it to be placed back on the banned substances list, while others argue that its usage has not changed since WADA removed it from its banned substances list.[8] It took most of the twentieth century for certain performance-enhancing substances to migrate to a cultural location where they were condemned as illegal technoscientific tools damaging the purity and authenticity of sport. But caffeine is a simple example of how these technoscientifically derived lists, which are far from immutable, will continue to evolve. These lists also reflect and represent the social and cultural fear of the invisible and unknown allied with the efficacious power that makes all forms of performance-enhancing drugs so problematic for sporting authenticity.

Internal, Maintains Authenticity

Technoscience residing in the upper-left-hand corner of the authenticity–body coordinate system is internal to the body but does not necessarily undermine the authenticity of sport. Examples of these technoscientific objects and practices can be seen in injury treatment. Very few elite athletes have injury-free careers. The most serious injuries require surgical repair and often the integration of foreign materials into the body. Techniques such as ulnar collateral ligament repair or materials such as titanium rods, screws, and plates represent a set of routine technologies and medical processes by which modern medicine repairs injured bodies.[9] Repair is key to understanding the difference between these internalized technoscientific practices and performance-enhancing drug usage. These technoscientific artifacts, like performance-enhancing drugs, cannot be seen by the naked eye; however, they do not produce the same level of vitriol about cheating and undermining the game seen with performance-enhancing drugs. These tools and techniques hopefully return a body to a normal, unenhanced, and pre-injury performance level.

The contrast is subtle but clear. Injury repair aims to restore a body to its prior state, or as close as modern medicine can make it, whereas performance-enhancing drugs unfairly enhance the "natural" abilities of a given body. Yet these boundaries can be very murky to a broader public. Seemingly miraculous recoveries by athletes such as NFL player Adrian Peterson after the surgical repair of a torn left anterior cruciate and medial collateral ligaments raise questions about which combination of technoscientific treatments athletes can and will use.[10] Conversely, one can argue that an injured body is the "natural" state, and repairing it to pre-injury levels of performance is an enhancement. This line of reasoning can feed into a morass of tautologous arguments about what is or is not the prior state or the natural limits of a human body. Yet questions about natural versus machine-augmented ability drive concerns about the coming place of prosthesis-using athletes competing in a host of sporting domains that will dwarf the concerns regarding Oscar Pistorius's Olympic runs. Nevertheless, publics and governing bodies generally have chosen to view the repair of an injured body as an appropriate and valuable use of career-saving technoscience.

External, Maintains Authenticity

Technoscience residing in the upper-right-hand corner neither undermines authenticity nor is hidden within or alters the body. Decision aids such as instant replay fit within this quadrant. Initially, many sports slowly implemented this technology, but it is now a familiar tool that can assist in refereeing increasingly fast-paced sporting competitions.[11] These types of technoscientific systems display their effectiveness when it is hard for the naked eye to make a verifiable judgment. These truth-making devices are not foolproof, and their utilization is far from uniform across sport.[12] What makes them unique and powerful is that they do not directly impinge on game play. They can and do cause longer-than-routine pauses in play, but slowing an athlete's or a team's momentum for closer and more accurate evaluation of play is a fair trade-off for most sports. Some governing bodies, such as the Fédération Internationale de Football Association, believe that human error is part of the game (though it embraced goal-line technology for the 2014 FIFA World Cup).[13] Similar evaluative technoscientific systems, such as Hawk-Eye, which produces an estimation of a ball's movement through space, have gained acceptance in tennis and cricket.[14] But decision aids, like all technoscientific interventions, produce unforeseen ripples and questions within sport. These types of evaluative technoscientific systems showed their limitations at the 2011 Track and Field World Championships in Daegu, South Korea.

In the 100 m final, the highly decorated sprinter Usain Bolt was the favorite, but he false started and was disqualified. At first glance, this would appear to be a simple case of one individual moving before others. The race's starter easily determined that Bolt was the offending runner because a sensor connected to the pressure-sensitive starting blocks provided a readout indicating which lane housed the offending runner. If this had happened to a less-well-known sprinter, or had occurred in an earlier round, this false start would not have been placed under such a high level of scrutiny. But Usain Bolt was not just any sprinter; he was attempting to become the greatest ever, so this race received a great deal of attention and analysis. For viewing audiences, it was easy to receive high-level feedback from a television or a stadium projection screen. These audiences quickly saw the replay showing Bolt's false start in lane 5, but it also showed another initially unseen action in lane 6. To Bolt's right was his countryman Yohan Blake. The slow-motion replay made it readily apparent that Blake twitched before the firing of the starting gun. In the interim, the starter called the runners back to the line, and the second attempt at the final went off without a hitch, minus the disqualified Bolt. Blake sped to the finish and became the youngest 100 m world champion.

In the days after the final, track and field blogs reverberated with discussions of Blake's twitch. The interest was so high that bloggers acquired the starting-line data.[15] The data showed that Bolt did indeed false start, but it also showed that Blake twitched, which historically may have merited disqualification. However, he did not surpass the threshold at which the electronic sensors in the starting blocks would register a false start. Electronic starting blocks were supposed to mitigate judgment errors. But would more skilled eyes, attuned to the rolls, wiggles, and twitches of elite sprinters, have detected the Blake twitch and disqualified him before Bolt? Perhaps, but the International Association of Athletics Federations solved this problem by legislatively solidifying the technoscientific power of false-starting equipment and sensors while simultaneously diminishing the knowledge and skill of the starter. According to Rule 162, "False Start," of the IAAF Competition Rules 2014–15, "an athlete, after assuming a full and final starting position, shall not commence his start until after receiving the report of the gun. If, in the judgement of the Starter or Recallers, he does so any earlier, it shall be deemed a false start." From these sentences, it appears that the starters can use their knowledge and experience to determine if an athlete false starts. Yet the subsequent sentences of Rule 162 make it clear that a technoscientific instrument actually determines false starts. "When an IAAF approved false start control apparatus is in use, the Starter and/or an assigned Recaller shall wear headphones in order to clearly hear the acoustic signal emitted when the apparatus indicates a possible false start (i.e. when the reaction time is less than 0.100 second). As soon as the Starter and/or an assigned Recaller hears the acoustic signal, and if the gun was fired, there shall be a recall and the Starter shall immediately examine the reaction times on the false start control apparatus in order to confirm which athlete(s) is/are responsible for the recall."[16]

The rule goes even further to address the Daegu final and reconfirms the primacy of the starting technoscience. For Blake's twitch, note (i) of Rule 162 indicates that "any motion by an athlete that does not include or result in the athlete's foot / feet losing contact with the foot plate(s) of the starting blocks, or the athlete's hand / hands losing contact with the ground, shall not be considered to be the commencement of his start." Thus the new rule allows for movement similar to that of Blake's twitch in the final with the penalty being that "such instances may, if applicable, be subject to a disciplinary warning or disqualification." This would seem to give a level of interpretive flexibility back to the starter, but note (iii) reconfirms that "when an IAAF approved false start control apparatus is in operation, the evidence of this equipment shall normally be accepted as conclusive by the Starter."[17] Thus the technoscientific decision aids always seem to trump human experience and knowledge. This question can be endlessly debated, but on the whole, sport governing bodies, publics, technoscientific actors, and athletes support these systems because they apparently assist in removing human error during critical moments, when the desire for clear refereeing and judging is high, while also trading human experience for assumed technoscientific accuracy.

External, Undermines Authenticity

The lower-right-hand quadrant is the space where technoscience is at a distance from the body and undermines the game or sport by taking power and authority away from a competitor. This realm comprises objects that athletes attach to their bodies, wear, ride, drive, and/or swing. In many ways, a significant portion of this book has focused on this quadrant because it is the site where the most contested and visible sport technoscientific innovation occurs. This is a space where devices and machines potentially supersede the athlete.

Historically, artifacts that produce this effect are eventually banned from a sporting competition. The continual negotiation of this form of tension is seen at the highest levels of multiple sporting arenas. For instance, automobile and motorcycle racing rely on the symbiotic relationship between the driver or rider and the machine. In motorsports racing, a key element of the sport's appeal is the machine and an athlete's ability to manage the machine's incredible power. Only the most skilled athletes can extract the best performance from these formidable devices. These sports evaluate the ability of the driver or rider based on how much performance he—and, more recently, she—can extract from the machine. Yet many racing formats struggle with the potency of computing, material science, and engineered aerodynamics. Traction control, telemetry, tire compounds, and driving simulators are a sample of the necessary technoscientific tools and equipment assisting drivers and teams in the management and control of exceedingly powerful vehicles.

Drivers or riders frequently speak of needing to find the right "package" to be competitive and win. This package is a mixture of human and nonhuman elements allowing a driver or rider access to the best fusion of machine and electronics. But this ability obfuscates another set of highly skilled actors who design, engineer, and fabricate these technological masterpieces. In these sporting competitions, it is becoming exceedingly difficult to deny the power of technoscience, so much so that it is becoming unclear whether the "real" competition is actually a technoscientific one.[18] In motorsports competition, when we cheer on a driver's mastery of a machine, should we cheer equally for those who brought the machine to life? Increasingly, the answer is yes. But what happens to the world of motorsport if this balance swings away from the drivers and more in the direction of the engineers or the machines?

In our contemporary sporting worlds, the comfortable and idealized place for technoscience is at the 0,0 point on the X and Y axis. At this point, technoscience is fully instrumental. It is understood to be benign and inconsequential material infrastructure of a game. By locating technoscience at 0,0, sport maintains a historical balance in which the body is the deciding factor within sporting competitions. But within modern sport, this equilibrium has become more difficult to hold. As the emphasis on and rewards for winning escalate, the demands to find an edge increase accordingly and compromise the noble vision that sporting competition is fair and that the best-trained bodies and minds win. This idealized vision has always been hard to sustain. Even most children find out very early in school physical education that the world of sport is a space in which fairness and equality are upheld loosely. Sport is about producing, promoting, and preserving inequalities. Athletes garner competitive advantages in many ways, but a large part of twentieth-century sport has been about gaining competitive edges through the use of technoscientific artifacts attached to, used by, or integrated into a body.

Many online discussions, televised debates, and courtroom battles about technoscience center on determining what is or is not permissible within a given sport. If one key goal is to maintain sport as a body-first endeavor, sport governing institutions, viewing publics, equipment manufacturers, and athletes have to, in varying degrees, participate in and contribute to the sport-technoscience dialectic. The examples presented in this book explore how these four constituencies attempt to deliberate, build, and support a tenuous balance between the historicized noble aims of sport and the advancing thrust of technoscience. But, like most social, cultural, and environmental equilibriums, they are difficult to maintain. By not fighting so hard to maintain the primacy of the athletic body within sport or believing that technoscience will solve all problems, sporting cultures can hopefully have a more proactive and integrative approach to new and emerging technoscience, rather than being fearful and reactionary. Finally, a look at the evolution of the bat in college baseball is instructive for understanding what is at stake when a sporting culture attempts to use the power of technoscience to reinstrumentalize a central object of a sport.

Reconfiguring the Baseball Bat

In baseball, the bat is central. One cannot play the game without a bat to strike the ball. The move from wooden to aluminum bats transformed college baseball in the United States. The introduction of aluminum bats around 1974 commenced a technoscientific evolution that transformed the way athletes played the game. Initially introduced as a potential cost-saving measure, the first versions of aluminum bats performed similar to wooden bats.[19] But manufacturers quickly realized that they could engineer higher-performing bats by harnessing aluminum's physical attributes. They made bats lighter so players could swing them faster. Thinner-walled aluminum in bat barrels also generated a trampoline effect, which enabled players to hit balls farther. In the highly competitive world of college baseball, a bat giving one player an advantage over his competitors became a highly desirable tool that undoubtedly influenced the outcome of games. Over the next decade, the performance capabilities of aluminum bats increased rapidly.

From 1974 to 1985, the National Collegiate Athletic Association's (NCAA) data on Division I baseball statistics show significant leaps in categories such as home runs, scoring, and batting averages, as well as a reciprocal decline in pitching earned-run averages and strikeouts.[20] During the same period, fielding percentages stayed roughly the same, so it can be deduced from the data that poor fielding was not responsible for the wholesale increase in offensive productivity during this ten-year span. Though a genetic bubble of gifted athletes or the increased professionalization and formalization of youth training could have been responsible for this escalation in batting efficiency, changes in the NCAA rules make a strong argument for aluminum alloy bats as a primary factor.

In 1985, the NCAA set a minimum weight limit for aluminum alloy bats to decrease the bat speed advantage that lightweight aluminum alloy bats provided. In the years following the implementation of this rule, offensive numbers fell. However, offensive numbers began to rise again quickly, and by the 1998 season Division I collegiate baseball teams collectively broke a massive number of offensive records. Bat manufacturers had once again learned how to extract the most performance from aluminum alloy bats by exploiting the material's malleability. In 1999, the NCAA developed a new set of rules to once again address the trampoline effect. Specifically, it created the ball exit speed ratio (BESR) standard to limit the speed at which a baseball could leave the bat barrel, instituted a reduction in the maximum bat barrel diameter to 2.625 inches, and enacted the minus-3 ratio, which meant that the difference in weight and length could not exceed three.[21] This worked fairly well until the 2009 season, when offensive numbers again charted uncomfortably upward, this time due to the adoption of reinforced carbon fiber polymer composite bats.[22] In a sense, the bat manufacturers applied their years of bat research to a new technoscientifically engineered material.

On July 17, 2009, the NCAA proposed "an immediate and indefinite moratorium on the use of composite barreled bats."[23] The NCAA was concerned most with the fact that "during the 2009 Division I Baseball Championship, 25 composite bats were selected for ball exit speed ratio (BESR) certification tests. Of the 25 bats, 20 failed the official BESR test for current NCAA performance levels."[24] These bats had been determined to have met the NCAA's BESR standard but now outperformed this standard. The NCAA implied that these bats had been systematically manipulated. The NCAA Baseball Rules Committee "began looking at the issue during the regular season after being alerted to the possible alteration of bats through a technique known as 'rolling.' In the rolling process, the bat is placed in a machine, where it is compressed. The process makes the bat softer and enhances the spring-like factor of the ball exiting off the bat's surface."[25] The chair of this committee, Bob Brontsema, stated that it is doubtful that rolling was a large-scale issue, but it was clear that "the performance of composite bats improves through repeated, normal use and [that] these bats often exceed acceptable levels."[26] It was after these revelations that the NCAA, like many other sport governing bodies before and after, chose a reactive response and banned composite bats.

In the never-ending quest to return the game to the athletes and reinstrumentalize the bat, the NCAA stepped in before the 2011–12 collegiate baseball season and again introduced a new standard: the batted ball coefficient of restitution (BBCOR),[27] which regularizes baseball bats by deadening the trampoline rebound that a pitched ball experiences. Its implementation led to a precipitous drop in offensive performance. Offensive numbers dropped into the range of the early 1970s, before the introduction of aluminum bats. Supporters of the new bats contend that it allows fans, recruiters, and trainers to truly see batting talent, skill, and ability. But by the end of the 2014 College Baseball season, not everyone was so happy with the way the new bat had changed the game. Reporting for *Sports Illustrated*, Jeff Bradley noted that in 2013 the number of home runs per game decreased by nearly one-third from the pre-BBCOR era, and that UCLA's 2013 national championship baseball team "tied a [College World Series] record by laying down a dozen sacrifice bunts in . . . five games."[28]

The latest crop of technoscientifically molded bats reshaped the play on the field by requiring a reconfiguration of winning strategy. Bradley argues that baseball returned to an older version of game play: small ball. Small ball emphasizes the advancing of runners through carefully orchestrated tactics. Small ball does not rely on home runs and aggressive batting. With the bats now limiting player power, coaches have resorted to small ball to win. The pendulum has swung so heavily toward small ball that players do not swing for the fences and scouts can no longer evaluate batting power or ability in game settings. Part of the reality of small ball is that universities evaluate coaches on winning at the collegiate level and not on the production of successful professional players. The BBCOR, designed to return hitting to a more pure and authentic state, influenced another set of demands not directly related to the noble aims of baseball, such as college coaches retaining their jobs. As a result, some coaches responded to the new standard that conceivably took away the importance of hitting by adjusting the way that they instructed their players to play the game of baseball. This rule, intended to give the game back to players, may have inadvertently taken more away from the players than high-performance bats did.[29]

If we read baseball bats of this period through a motor-over-machine allegory, one can conclude that since baseball is such a difficult sport to master, aluminum or composite bats, though they may improve a poor player's performance, cannot supersede the physiological talents of a "truly" gifted player. Though aspects of this reasoning are valid, diminishing the bat's technoscientific power and championing the body's ability does not alter the fact that bat performance can change the game. The power, historical value, and use of the motor-over-machine perspective demand the elevation of the body as a response to technoscientific innovation. This maneuver problematically sublimates technoscience to body but, more importantly, misses an entire set of equally important human actors in this dialectic—those who design the equipment. In the market-driven world of baseball bat production, economic gain and technical curiosity can be strong motivators for engineers, physicists, and designers to take baseball bats to illogical high-performance extremes. Those making bats or any technoscientific sporting equipment are competing just as vigorously as the athletes on the field, and, as such, sporting competitions cannot only be seen narrowly through athletic bodies. Sporting competitions must be understood through the interplay of morphing networks of human and nonhuman factors, many of which are far away from a competitive playing field. In this regard, it is valuable to probe how the power of motor-over-machine phrases not only uphold an idealized authenticity within sporting narratives but also support a set of commercial, financial, and professional aims encompassing modern sport.

What is interesting about the evolution of these new standards is that the NCAA attempted to erase the technoscientific power of the bats by reinstrumentalizing them as mundane pieces of equipment. Though it can be argued that aluminum revealed that a bat was never a simple or instrumental component of the game, the mere creation of the BESR and BBCOR standards provides undeniable evidence of the extent to which bats, as technoscientific objects, impact baseball. Furthermore, these series of regulations do not seem to have achieved the NCAA's desired outcome of reducing the importance of the bats—they just made the bats important in a different way, and seem to have made them even more meaningful. Each rule change did little to resolve the bat problem or, more broadly, make the technoscience in sport less problematic. The BESR and BBCOR standards return baseball to the motor-over-machine debate, but, more directly, the standards reinforce the collective illusion that in athletic competitions bodies should be the final arbiters of who wins and loses and not part of a loose and messy network of human, institutional, and material forces that shape the final results of athletic competitions. It is this historically and culturally rooted deception that allows us, no matter how much it may change the games people play, to continually relegate, force, or legislate technoscience into a marginalized position of instrumental sporting equipment.

Cultures, Technoscience, and Sporting Futures

As the history of technoscience and sport continually reveals, it is exceedingly less likely that sporting cultures will naively construct, consume, evaluate, and assess the "pure" athletic body as they have in the past. As we gather a better understanding of what bodies can and cannot do "naturally," sporting cultures are no longer yearning for the next great athletic performances in the way they once did. Great leaps in performance are no longer looked at with awe and admiration but are examined critically with skeptical eyes. This mistrust is well earned. In sporting worlds where there is a great deal to gain psychologically and financially by winning, athletes will always look for a competitive edge.

Sadly, dominant narratives about gains in athletic performance are tied intimately to the history of the rise of performance-enhancing drugs. As a result, sport existed in a "drug" era for most of the twentieth century. Recent public admissions from champions such as Lance Armstrong and Marion Jones support efforts to climb out of this woeful history, but the transition is far from over. This clearing of the air is undoubtedly vital for the future of all sport, but the focus on performance-enhancing drugs also obfuscates many other forms of performance enhancement. Most importantly, it overlooks the epic evolutions in chemistry, material science, engineering, design, and a host of other technoscientific areas that have revolutionized training, equipment, and the play of sport, and proves that sport has always been a technoscientific enterprise. This oversight is surprising because in the motor-over-machine dialectic, the body is described as a machine by defining it as a motor. Though not hard-wired to understand the body through mechanistic analogies, medical technoscience probes and prods the body to conceptually construct it as a rich matrix of mechanical interactions. With mechanistic metaphors replete within contemporary sport, it is striking that material technoscience has not received substantive analysis. Just as athletes administer pharmaceutical and nonpharmaceutical treatments ranging from coca leaves to strychnine to anabolic steroids, they also uncover new equipment, devices, and training techniques to provide the largest advantages—legal or illegal—over their competition.

The production of scientific knowledge and technological instruments is a social process embedded within an ever-changing network of cultural interactions. The idea of getting a better, truthful, or more authentic athletic performance by increasing the scientific and technological oversight is fraught with problems. This reality has troubling ramifications for sport. Thus how do we find a middle ground to understand, evaluate, and critique athletic performances mediated by emerging technoscience and its newly developed diagnostic instruments? It appears that in the near future, sport will be just as much about tests of human endurance and performance as it will be about testing the efficacy of technoscientific knowledge, diagnostic protocols, and material artifacts. It is the impact of the material artifacts and technoscientific evaluations that this book explores in an effort to understand what is at stake for the future of sport. Collectively, these examples challenge sporting cultures to rethink the primacy of the body in the narrative of athletic performance and to reconceptualize athletic performance as an evolving network of bodies and material artifacts that is as much a human endeavor as it is a technoscientific enterprise. In a sense, we must abandon the motor-over-machine argument and move toward understanding the symbiotic relationships between the motor *and* the machine, the body and the artifact. But before we can make the leap, we must more fully understand the cultures of sport and the multiple social, cultural, and institutional affordances that delimit this step.

Notes

Introduction. Sports, Bodies, and Technoscience

1. A colleague, who is also a tennis player, has spoken to me about using the phrase "it is the magician, not the wand" to motive his own play in a similar way.

2. The track was renamed the Sir Roger Bannister athletics track on May 10, 2007, in honor of his achievement.

3. Leo Marx, *The Machine in the Garden: Technology and the Pastoral Ideal in America* (New York: Oxford University Press, 1964).

4. Lewis Mumford, *Technics and Civilization* (New York: Harcourt, Brace, 1934); Michael Pollan, *The Omnivore's Dilemma: A Natural History of Four Meals* (New York: Penguin Press, 2006).

5. Franz Konstantin Fuss, Aleksandar Subic, and Sadayuki Ujihashi, eds., *The Impact of Technology on Sport II* (London: Taylor and Francis, 2008).

6. George Basalla, *The Evolution of Technology* (New York: Cambridge University Press, 1988).

7. Donald A. MacKenzie and Judy Wajcman, eds., *The Social Shaping of Technology*, Second Edition (Philadelphia: Open University Press, 1999).

8. Richard Sennett, *The Craftsman* (New Haven, CT: Yale University Press, 2008).

9. Siva Vaidhyanathan, *The Googlization of Everything: (And Why We Should Worry)* (Berkeley: University of California Press, 2011).

10. Ivo van Hilvoorde, Rein Vos, and Guido de Wert, "Flopping, Klapping, and Gene Doping: Dichotomies between 'Natural' and 'Artificial' in Elite Sport," *Social Studies of Science* 37 (2007): 173–200.

11. Howard Bryant, "The Truth," *ESPN The Magazine*, December 12, 2011, p. 10.

12. Bruno Latour, *Pandora's Hope: Essays on the Reality of Science Studies* (Cambridge, MA: Harvard University, 1999); Roslyn Kerr, "Assembling High Performance: An Actor Network Theory Account of Gymnastics in New Zealand," PhD diss., University of Canterbury, 2010.

13. Douglas B. Holt, John A. Quelch, and Earl L. Taylor, "How Global Brands Compete," *Harvard Business Review*, 82 (September 2004): 68–75.

14. Ronald R. Kline, "Where Are the Cyborgs in Cybernetics?" *Social Studies of Science* 39, no. 3 (2009): 331–62; Ray Kurzweil, *The Singularity Is Near: When Humans Transcend Biology* (New York: Penguin Books, 2006); Andrew Pickering, *The Cybernetic Brain: Sketches of Another Future* (Chicago: University of Chicago Press, 2010).

15. Thomas Hobbes, *Leviathan*, edited with an introduction and notes by J. C. A. Gaskin (Oxford: Oxford University Press, 1998), 7.

16. Stephen Jay Gould, *The Mismeasure of Man* (New York: Norton, 1981); Gavin Schaffer, "'Scientific' Racism Again? Reginald Gates, the Mankind Quarterly and the

Question of 'Race' in Science after the Second World War," *Journal of American Studies* 41, no. 2 (2007): 253–278; William H. Tucker, *The Funding of Scientific Racism: Wickliffe Draper and the Pioneer Fund* (Urbana: University of Illinois Press, 2007); William H. Tucker, *The Science and Politics of Racial Research* (Urbana: University of Illinois Press, 1994); Harriet A. Washington, *Medical Apartheid: The Dark History of Medical Experimentation on Black Americans from Colonial Times to the Present* (New York: Anchor Books, 2007).

17. Donna J. Haraway, *Simians, Cyborgs, and Women: The Reinvention of Nature* (New York: Routledge, 1991); N. Katherine Hayles, *How We Became Posthuman: Virtual Bodies in Cybernetics, Literature, and Informatics* (Chicago: University of Chicago Press, 1999).

18. Jan Rintala, "Sport and Technology: Human Questions in a World of Machines," *Journal of Sport and Social Issues* 19, no. 1 (1995): 62–75.

19. Dave Dravecky and Mike Yorkey, *Called Up: Stories of Life and Faith from the Great Game of Baseball* (Grand Rapids, MI: Zondervan, 2004), 138–139.

20. Sigmund Loland, *Fair Play in Sport: A Moral Norm System* (New York: Routledge, 2002), 41–106.

21. Erin E. Floyd, "The Modern Athlete: Natural Athletic Ability or Technology at Its Best," *Villanova Sports and Entertainment Law Journal* 9, no. 1 (2002): 155–180.

22. Ian Woodward, "Material Culture and Narrative: Fusing Myth, Materiality, and Meaning," in *Material Culture and Technology in Everyday Life: Ethnographic Approaches*, ed. Phillip Vannini (New York: Peter Lang, 2009), 60.

23. Lisa Rosner, *The Technological Fix: How People Use Technology to Create and Solve Problems* (New York: Routledge, 2004).

24. Woodward, "Material Culture and Narrative," 62.

25. Sigmund Loland, "Technology in Sport: Three Ideal-Typical Views and Their Implications," *European Journal of Sport Science* 2, no. 1 (2002): 1–11.

26. Brandon Richard, "Here's the Gold Shoes Usain Bolt Is Wearing to Dominate the Olympics," *Sole Collector*, August 17, 2016, http://solecollector.com/news/2016/08/usain-bolt-gold-puma-olympic-spikes; "New Mondotrack WS Track Surface to Serve as Competition Surface of 2016 Rio Olympic Games," *NASDAQ GlobeNewswire*, August, 25, 2015, https://globenewswire.com/news-release/2015/08/25/763152/10147028/en/New-Mondotrack-WS-track-surface-to-serve-as-competition-surface-of-2016-Rio-Olympic-Games.html.

27. Tara Magdalinski, *Sport, Technology, and the Body: The Nature of Performance* (New York: Routledge, 2009), 6.

28. Lanty M. O'Connor and John A. Vozenilek, "Is it the Athlete or the Equipment? An Analysis of the Top Swim Performances from 1990 to 2010," *Journal of Strength and Conditioning Research* 25, no. 12 (December 2011): 3239–3241.

29. Benjamin S. Roberts, Khaled S. Kamel, Clay E. Hedrick, Scott P. McLean, and Rick L. Sharp, "Effect of a FastSkin Suit on Submaximal Freestyle Swimming," *Medicine and Science in Sports and Exercise* 35, no. 3 (March 2003): 519–524.

30. Geoffroy Berthelot, Stéphane Len, Philippe Hellard, Muriel Tafflet, Nour El Helou, Sylvie Escolano, Marion Guillaume, Karine Schaal, Hala Nassif, François Denis Desgorces, and Jean-François Toussaint, "Technology and Swimming: 3 Steps beyond Physiology," *Materials Today* 13, no. 11 (2010): 46–51.

31. On November 30, 2011, Speedo released its FASTSKIN3 system, arguing once again that its technology would provide an increased level of medal-winning swimming efficiency. "Speedo Unveils FASTSKIN," Speedo International Limited, November 30, 2011, https://web.archive.org/web/20111204155410/http://newsroom.speedo.com /speedo-news/speedo-unveils-fastskin3/.

32. Kenny Moore, *Bowerman and the Men of Oregon: The Story of Oregon's Legendary Coach and Nike's Cofounder* (Emmaus, PA: Rodale Books, 2006); Rachel Bachman, "Nike's Holy Grail: Bowerman Family Unearths Long-Lost Waffle Iron," February 28, 2011, *The Oregonian*, http://blog.oregonlive.com/behinducksbeat/2011/02/nikes_holy_grail _bowerman_fami.html.

33. Union Cycliste Internationale, "UCI Hour Record," September, 9, 2000, https://web.archive.org/web/20001030005240/http://www.uci.ch/english/news/comm _20000908.htm.

34. Lydia Polgreen and Alan Cowell, "Pistorius Is Indicted on Murder Charge," *New York Times*, August 19, 2013, http://www.nytimes.com/2013/08/20/world/africa/oscar -pistorius.html.

35. Patricia J. Zettler, "Is It Cheating to Use Cheetahs? The Implications of Technologically Innovative Prostheses for Sports Values and Rules," *Boston University International Law Journal* 27, no. 2 (2009): 367–409.

36. "IAAF Council Introduces Rule Regarding 'Technical Aids,'" Press Release, March 26, 2007, http://www.iaaf.org/news/news/iaaf-council-introduces-rule -regarding-techni.

37. Jeré Longman, "An Amputee Sprinter: Is He Disabled or Too-Abled?" *New York Times*, May 15, 2007, http://www.nytimes.com/2007/05/15/sports/othersports/15runner .html?pagewanted=all&_r=0; Josh McHugh, "Blade Runner," *Wired*, March 2007, http://archive.wired.com/wired/archive/15.03/blade.html.

38. "Arbitration CAS 2008/A/1480 Pistorius v/ IAAF," *Australian and New Zealand Sports Law Journal* 145, no. 3 (2008): 145–164, http://www.austlii.edu.au/au/journals /ANZSportsLawJl/2008/7.html.

39. Ibid.

40. Randall Mayes, "The Modern Olympics and Post-Modern Athletics: A Clash in Values," *The Journal of Philosophy, Science and Law* 10 (2010): 1–17.

41. Kaye N. Ballantyne, Manfred Kayser, and J. Anton Grootegoed, "Sex and Gender Issues in Competitive Sports: Investigation of a Historical Case Leads to a New Viewpoint," *British Journal of Sports Medicine* 46, no. 8 (2012): 614–617.

42. Bennett Foddy and Julian Savulescu, "Time to Re-evaluate Gender Segregation in Athletics?" *British Journal of Sports Medicine* 45, no. 15 (December 2010): 1184–1188.

43. Richard Klaus Müller, "History of Doping and Doping Control," in *Handbook of Experimental Pharmacology: Doping in Sport*, ed. Detlef Thieme and Peter Hemmersbach (New York: Springer, 2010), 1–23.

44. Larry D. Bowers, "The International Antidoping System and Why It Works," *Clinical Chemistry* 55, no. 8 (2009): 1456–1461.

45. "Extras from the Morning Show with Boomer Esiason and Craig Carton," September 25, 2012, http://nyc.podcast.play.it/media/do/do/d1/d1/dC/d3/dB/11C3B_3 .MP3.

46. Mario Zorzoli and Francesca Rossi, "Implementation of the Biological Passport: The Experience of the International Cycling Union," *Drug Testing and Analysis* 2, nos. 11–12 (2010): 542–547.

47. Giuseppe Lippi, Mario Plebani, Fabian Sanchis-Gomar, Giuseppe Banfi, "Current Limitations and Future Perspectives of the Athlete Blood Passport," *European Journal of Applied Physiology* 112, no. 10 (2012): 3693–3694.

48. Daniel Benson, "Are Micro-dosing Riders Poking Holes in Biological Passport?" *CyclingNews*, May 6, 2015, http://www.cyclingnews.com/features/are-micro-dosing -riders-poking-holes-in-biological-passport/.

Chapter 1. Black Is the New Fast

1. Andrew Dampf, "International Capsules: A Swimming Stunner: Phelps Loses to Biedermann," *The Brownsville Herald*, July 28, 2009, http://www.brownsvilleherald.com /sports/article_05788681-6433-58eb-90ca-202dd5f168cc.html.

2. Maggie Barry, "Forgotten Olympic Golden Girl Belle Moore Remembered 100 Years After Landmark Win," *Daily Record and Sunday Mail*, April 29, 2012, http:// www.dailyrecord.co.uk/news/uk-world-news/forgotten-olympic-golden-girl-belle -1120444.

3. Paul Wilson, *Guide to the Archive of Speedo* (Sydney: Powerhouse Museum, 2006), 1. The company did not abandon underwear completely. In fact, in 1937 they signed a licensing agreement with American company Coopers, Inc., and eventually began to manufacture and sell Jockey underwear.

4. Tamás László, "Alfréd Hajós, the Hungarian Dolphin," *Hajós Alfréd Society*, https://web.archive.org/web/20070801000000*/http://www.hajosalfred.hu/data /AlfredHajos_Aphrodisias2005.doc.

5. Tiago M. Barbosa, José A. Bragada, Víctor M. Reis, Daniel A. Marinho, Carlos Carvalho, and António J. Silva, "Energetics and Biomechanics as Determining Factors of Swimming Performance: Updating the State of the Art," *Journal of Science and Medicine in Sport* 13, no. 2 (March 2010): 262–269.

6. Michael Sheard and Jim Golby, "Effect of a Psychological Skills Training Program on Swimming Performance and Positive Psychological Development," *International Journal of Sport and Exercise Psychology* 4, no. 2 (2006): 149–169.

7. Joseph Clois Shivers Jr., United States Letters Patent #3,044,989, filed August 5, 1958, patented July 17, 1962.

8. Kaori O'Connor, *Lycra: How a Fiber Shaped America* (New York: Routledge, 2011): 54–83.

9. Wilson, *Guide to the Archive of Speedo*, 2.

10. Hideki Takagi, "A Current of Product Development for Competitive Swimsuits," *Japanese Society of Mechanical Engineers News* 15, no. 2 (December 2004): 8.

11. Patrice A. Keats and William R. Keats-Osborn, "Overexposed: Capturing a Secret Side of Sports Photography," *International Review for the Sociology of Sport* 48, no. 6 (2012): 643–657.

12. Christine Schmidt, *The Swimsuit: Fashion from Poolside to Catwalk* (New York: Bloomsbury Academic, 2012).

13. J. C. Chatard, J. M. Lavoie, B. Bourgoin, and J. R. Lacour, "The Contribution of Passive Drag as a Determinant of Swimming Performance," *International Journal of Sports Medicine* 11, no. 5 (1990): 367–372.

14. Yohei Sato and Takanori Hino, "CFD Simulation of Flows around a Swimmer in a Prone Glide Position," *Japanese Journal of Sciences in Swimming and Water Exercise* 13, no. 1 (2010): 1–9.

15. T. Togashi, T. Nomura, and M. Fujimoto, "A Study of Low Resistance Swimming Suit for Competitive Swimming," *Descent Sports Science* 10 (1989): 75–82.

16. Takagi, "A Current of Product Development for Competitive Swimsuits," 8.

17. Yukimaru Shimizu, Toshiaki Kuzuki, Kunihito Suzuki, and Hiroshi Kiyokaya, "Studies on Fluid Drag Measurement and Fluid Drag Reduction of Woman Athlete Swimming Suit," *Transactions of the Japan Society of Mechanical Engineers* 63, no. 616 (1997): 3921–3927.

18. Andrew D. Grainger, Joshua I. Newman, and David L. Andrews, "Global Adidas: Sport, Celebrity, and the Marketing of Difference," in *Global Sport Sponsorship*, ed. John M. Amis and T. Bettina Cornwell (New York: Berg, 2005): 89–105.

19. Greg Hunter, *Ian Thorpe: The Biography* (Sydney: Macmillan, 2004).

20. "Adidas Presents New Bodysuit: The JETCONCEPT," July 19, 2003, http://www .eurekalert.org/pub_releases/2003-07/aa-apn071803.php. The JETCONCEPT suit was not the first use of riblets in sporting competition. A "riblet skin" was applied to United States rowing shells for the 1984 Los Angeles Olympics and the *Stars and Stripes 88 (US-1)* in the 1988 America's Cup. "NASA Goes to the Olympics," August 12, 2004, http://www.nasa.gov/audience/forstudents/5-8/features/F_NASA_Goes_to_the _Olympics.html.

21. Ronald D. Joslin, "Aircraft Laminar Flow Control," *Annual Review of Fluid Mechanics* 30 (1998): 1–29.

22. Takagi, "A Current of Product Development for Competitive Swimsuits," 8.

23. "SPEEDO Introduces Fastskin—the Fastest Swimsuit Ever Made," *Business Wire*, March 16, 2000, http://www.thefreelibrary.com/SPEEDO+Introduces+Fastskin+—+the +Fastest+Swimsuit+Ever+Made.-a060827273.

24. "Arena Powerskin R-Revolution," March 18, 2008, http://www.sfilate.it/14637 /arena-powerskin-r-evolution.

25. Brent S. Rushall, "Swimming Bodysuits: A Violation of FINA Rule SW 10.7," April 2, 2000, http://coachsci.sdsu.edu/swim/bodysuit/CASsub.htm.

26. Ibid.

27. N. Benjanuvatra, G. Dawson, B. A. Blanksby, and B. C. Elliott, "Comparison of Buoyancy, Passive and Net Active Drag Forces between Fastskin and Standard Swimsuits," *Journal of Science and Medicine in Sport* 5, no. 2 (2002): 115–123.

28. Ibid., 122.

29. Rick L. Sharp and David L. Costill, "Influence of Body Hair Removal on Physiological Responses during Breaststroke Swimming," *Medicine and Science in Sports and Exercise* 21, no. 5 (1989): 576–580.

30. J. P. Clarys, "Human Body Dimensions and Applied Hydrodynamics: Selection Criteria for Top Swimmers," *SNIPES Journal* 9 (1986): 32–41.

31. Huub M. Toussaint, Martin Truijens, Meint-Jan Elzinga, Ad van de Ven, Henk de Best, Bart Snabel, and Gert de Groot, "Effect of a Fast-Skin 'Body' Suit on Drag during Front Crawl Swimming," *Sport Biomechanics* 1, no. 1 (2002): 1–10.

32. Ibid., 9.

33. Ibid., 8.

34. Jennifer Craik, "The Fastskin Revolution: From Human Fish to Swimming Androids," *Culture Unbound: Journal of Current Cultural Research* 3, no. 1 (2011): 75.

35. Donna Haraway, "A Manifesto for Cyborgs, Science, Technology, and Socialist Feminism in the 1980s," *Australian Feminist Studies* 2, no. 4 (1987): 1–42.

36. "The Powerskin History," https://web.archive.org/web/20121025015752 /http://www.arenapowerskin.com/the-powerskin-history/.

37. Frank Vizard, "The Olympian's New Clothes: High Tech Apparel May Determine who Takes Home the Gold," *Scientific American*, August 9, 2004, http://www.scientific american.com/article.cfm?id=the-olympians-new-clothes.

38. Ibid.

39. Ibid.

40. Ibid.

41. Joseph C. Mollendorf, Albert C. Termin II, Eric Oppenheim, and David Pendergast, "Effect of Swim Suit Design on Passive Drag," *Medicine and Science in Sports and Exercise* 36, no. 6 (June 2004): 1029–1035.

42. Sean Parnell, "Slippery Business," *The Australian*, May 31, 2008, http://www .theaustralian.com.au/news/features/slippery-business/story-e6frg8h6 -1111116490096.

43. Ben Doherty, "Fast Swimsuit 'Perfectly Legal': Tester," *The Age*, March 28, 2008, http://www.theage.com.au/news/swimming/fast-swimsuit-perfectly-legal-tester/2008 /03/25/1206207106395.html; Barry Bixler, "Engineering Design and Analysis of the Speedo FSII Swimsuit," May 16, 2006, http://events.stanford.edu/events/87/8736/; Benjamin S. Roberts, Khaled S. Kamel, Clay E. Hedrick, Scott P. McLean, and Rick L. Sharp, "Effect of a FastSkin Suit on Submaximal Freestyle Swimming," *Medicine and Science in Sports and Exercise* 35, no. 3 (March 2003): 519–524; Kathy Barnstorff, "Olympic Swimmers Shattering Records in NASA-Tested Suit," August 13, 2008, http://www.nasa.gov/topics/technology/features/2008-0813-swimsuit.html.

44. Steve Wilkinson, quoted in Barnstorff, "Olympic Swimmers Shattering Records in NASA-Tested Suit."

45. Parnell, "Slippery Business."

46. Ernest Beck, "Speedo: Innovation in the AquaLab," *Bloomberg Businessweek*, April 14, 2008, http://www.businessweek.com/stories/2008-04-14/speedo-innovation -in-the-aqua-labbusinessweek-business-news-stock-market-and-financial-advice.

47. Ibid.

48. Parnell, "Slippery Business."

49. Eric Wilson, "Swimsuit for the Olympics Is a New Skin for the Big Dip," *The New York Times*, February 13, 2008, http://www.nytimes.com/2008/02/13/sports/othersports /13swim.html?scp=2&sq=Swimsuit+for+the+Olympics+Is+a+New+Skin+for+the+Big +Dip&st=nyt.

50. "SPEEDO Unveils Revolutionary Elite Speed Suit: The LZR RACER," *Business Wire*, February 12, 2008, http://www.businesswire.com/news/home/20080212006580 /en/SPEEDO-Unveils-Revolutionary-Elite-Speed-Suit-LZR.

51. "FINA Rules to Allow High-Tech Swimsuits; Italian Coach Calls It 'Technological Doping,'" *The New York Times*, April 8, 2008, http://www.nytimes.com/2008/04/08 /sports/08iht-swim8.11783912.html.

52. "Space Age Swimsuit Reduces Drag, Breaks Records," *NASA Spinoff*, accessed October 12, 2015, http://spinoff.nasa.gov/Spinoff2008/ch_4.html.

53. "Further Swimsuit Developments," *The Science of Sport*, April 8, 2008, http:// sportsscientists.com/2008/04/further-swimsuit-developments/.

54. "Arena Calls for 'Urgent' Testing of Swimsuit," *The Telegraph*, April 11, 2008, http://www.telegraph.co.uk/sport/olympics/swimming/2297031/Arena-calls-for -urgent-testing-of-swimsuit.html.

55. "Olympic Swimmers Face Tough Choices after Record-Breaking Speedo Suit Receives FINA Backing," *The New York Times*, April 13, 2008, http://www.nytimes.com /2008/04/13/sports/13iht-swimmingsuit13.11939623.html?_r=0.

56. Martin Petty, "FINA Pressed to Settle Row over High-Tech Swimsuits," *Reuters*, April 11, 2008, http://www.reuters.com/article/2008/04/11/us-swimming-world -bodysuits-idUSSP20265420080411.

57. "PR18 - FINA Press Release on Arena Open Letter to FINA on Swimsuits," *FINA*, April 6, 2008, http://www.fina.org/news/pr18-fina-press-release-arena-open-letter-fina -swimsuits.

58. "PR21 - Meeting FINA/Swimwear Manufacturers," *FINA*, April 12, 2008, http://www.fina.org/news/pr21-meeting-finaswimwear-manufacturers.

59. Craig Lord, "Schubert: Clarity in the Costume Drama," *SwimNews.com*, April 8, 2008, http://swimnews.com/News/view/6028.

60. Ibid.

61. Nick Pearce, "Olympic Swimming Records Set to Tumble at London 2012 as Speedo Unveil Fastskin3 Swimwear System," *The Telegraph*, November 30, 2011, http:// www.telegraph.co.uk/sport/olympics/swimming/8924083/Olympic-swimming-records -set-to-tumble-at-London-2012-as-Speedo-unveil-Fastskin3-swimwear-system.html.

62. "PR85 - FINA Information on the Update of the Swimwear Approval Proce- dures," *FINA*, December 22, 2008," http://www.fina.org/news/pr85-fina-information -update-swimwear-approval-procedures.

63. "FINA Sets March Date for Action on High-Tech Suits," *Teamusa.org*, Decem- ber 22, 2008, http://www.teamusa.org/News/2008/December/22/FINA-sets-March -date-for-action-on-high-tech-suits.

64. "Britain Backs Protest over LZR Racer," *The Guardian*, December 15, 2008, http://www.theguardian.com/sport/2008/dec/15/swimming.

65. "PR16 - Meeting between FINA and Swimwear Manufacturers," *FINA*, Febru- ary 20, 2009, http://www.fina.org/news/pr16-meeting-between-fina-and-swimwear -manufacturers.

66. "Dubai Charter on FINA Requirements for Swimwear Approval," PR21 - FINA Bureau Meeting, March 14, 2009, http://www.fina.org/news/pr21-fina-bureau-meeting.

67. Ibid.

68. "PR37 - FINA Commission for Swimwear Approval," *FINA*, May 19, 2009, http://www.fina.org/news/pr37-fina-commission-swimwear-approval.

69. "PR46 - FINA Executive Meeting / Swimsuits," *FINA*, June 22, 2009, http://www.fina.org/news/pr46-fina-executive-meeting-swimsuits.

70. Ibid.

71. "Confirmation of the Approval for Swimsuit J01 Created and Manufactured by Jaked Srl, Official Technical Sponsor and Official Supplier of the Italian Teams of FIN (Italian Swimming Federation)," *Jaked*, June 22, 2009. http://www.dimnp.unipi.it/lanzetta/images/jakeden.pdf.

72. *Arbitration CAS 2009/A/1917 Amaury Leveaux and Aurore Mongel v. Fédération Internationale de Natation (FINA)*, order of 29 July 2009.

73. "FINA Upholds New Swimsuit Limits," *Associated Press*, July 28, 2009, http://sports.espn.go.com/oly/swimming/news/story?id=4361323.

74. "PR58 - FINA Bureau Meeting," *FINA*, July 28, 2009, http://www.fina.org/news/pr58-fina-bureau-meeting.

75. Ibid.

76. Ibid.

77. "PR59 - FINA Bureau Meeting," *FINA*, July 31, 2009, http://www.fina.org/news/pr59-fina-bureau-meeting.

78. "Rio 2016—How Many Swimming World Records Should We Expect?," *The Stats Zone*, August 4, 2016, http://www.thestatszone.com/articles/rio-2016-how-many-swimming-world-records-should-we-expect.

Chapter 2. Gearing Up for the Game

1. Clifton Brown, "GOLF; Callaway Is Prepared to Fight Any Ban on Clubs," *The New York Times*, May 23, 1998, http://www.nytimes.com/1998/05/23/sports/golf-callaway-is-prepared-to-fight-any-ban-on-clubs.html; Federation Internationale de l'Automobile, "2017 Formula One Technical Regulations," http://www.fia.com/file/40961/download/14591?token=5zH5vGzI.

2. Robert Imre, "Hungary's Revolutionary Golden Team," in *Soccer and Philosophy: Beautiful Thoughts on the Beautiful Game*, ed. Ted Richards (Chicago: Open Court, 2010), 290–301.

3. Arthur Heinrich, "The 1954 Soccer World Cup and the Federal Republic of Germany's Self-discovery," *American Behavioral Scientist* 46, no. 11 (2003): 1491–1505.

4. Barbara Smit, *Sneaker Wars: The Enemy Brothers Who Founded Adidas and Puma and the Family Feud that Forever Changed the Business of Sport* (New York: Harper Perennial, 2009), 1–14.

5. It should be noted that his brother Rudolf started a competing company initially named "Ruda" but would shortly change its name to "Puma."

6. Barbara Smit, *Pitch Invasion: Three Stripes, Two Brothers, One Feud: Adidas, Puma and the Making of Modern Sport* (New York: Penguin, 2006), 45–58.

7. James Roberts, "A Welshman's World Cup Final," June 10, 2010, http://www.bbc.co.uk/blogs/waleshistory/2010/06/wales_world_cup_final_mervyn_griffiths.html.

8. Kate Connolly, "World Champions or Soccer Cheats?" *The Telegraph*, April 1, 2004, http://www.telegraph.co.uk/news/worldnews/europe/germany/1458294/World -champions-or-soccer-cheats.html.

9. Marvin Barias, "The True Story behind the Banned Air Jordan," *Sole Collector*, July 21, 2016, http://solecollector.com/news/2016/07/the-true-story-behind-the -banned-air-jordan#ixzz3GgNFx3ZH.

10. Faruq Gbadamosi, "The Economics of Nike's Air Jordan Brand," *The Market Mogul*, September, 19, 2015, http://themarketmogul.com/the-economics-of-nikes-air -jordan-brand/.

11. Barias, "The True Story behind the Banned Air Jordan."

12. Ibid.

13. Pete Forester, "How One of the Most Iconic Sneakers in History Almost Didn't Happen," *Esquire*, July 20, 2016, http://www.esquire.com/style/news/a46831/air-jordan -1-history/.

14. "Michael Jordan 'Banned' Commercial," https://youtu.be/zkXkrSLe-nQ.

15. Forester, "How One of the Most Iconic Sneakers in History Almost Didn't Happen."

16. "Nike Basketball 1989—'It's Gotta Be The Shoes—Commercial," https://youtu.be /Abr_LU822rQ.

17. "National Basketball Association (NBA) Bans APL Shoes," October 18, 2010, https://www.athleticpropulsionlabs.com/index.php/news/detail/index/id/14/.

18. Ibid.

19. United States Letter Patent #8,112,905 B2, "Forefoot Catapult for Athletic Shoes," filed May 18, 2009, issued February 12, 2012.

20. Carla Lazzareschi, "More Change in Top Ranks of L.A. Gear," *Los Angeles Times*, September 27, 1991, http://articles.latimes.com/1991-09-27/business/fi-3001_1_chief -operating-officer.

21. Chris Ballard, "Can Special Basketball Shoes Really Make You Jump Higher," *Sports Illustrated*, September 8, 2010, http://www.si.com/more-sports/2010/09/08 /shoes.

22. *Official Rules of the National Basketball Association, 2012–2013*, p. 10, http://www .nba.com/media/dleague/2012-13_NBA_RULE_BOOK.pdf.

23. Abram Brown, "How an NBA Ban Fueled Our Business," *Inc.*, December 8, 2011, http://www.inc.com/abram-brown/how-an-nba-ban-fueled-our-business.html.

24. "Load N' Launch Technology," accessed October 13, 2015, http://www .athleticpropulsionlabs.com/load-n-launch-technology.html.

25. See Ballard, "Can Special Basketball Shoes Really Make You Jump Higher"; and Kevin Hung, "APL Concept 1 Review," *Counterkicks.com*, September 15, 2010, https://web .archive.org/web/20120804205846/http://counterkicks.com/2010/09/apl-concept-1 -review-by-kevin-hung/.

26. "Illegal Shoe," December 6, 2010, *Officiating.com Blog*, http://forum.officiating .com/basketball/60020-illegal-shoe.html#post705609.

27. "NFHS Ban On Concept 1 Shoes Is Lifted," Alabama High School Athletic Association, February 3, 2011, http://dnn.ahsaa.com/AHSAA/News/NFHSBanOn Concept1ShoesIsLifted/tabid/3323/Default.aspx.

28. Shana Renee, "Update! Adidas Announces the Release of the adiZero Crazy Light," *All Sports Everything*, April 14, 2011, http://allsportseverything.com/2011/04 /hours-away-from-adidas-light-as-_-_-_-_-unveiling/.

29. Ibid.

30. Torsten Brauner, Marc Zwinzscher, and Thorsten Sterzing, "Basketball Footwear Requirements are Dependent on Playing Position," *Footwear Science* 4, no. 3 (2012): 191–198.

31. Robert L. Ashford, Peter White, Clive E. Neal-Sturgess, Nachiappan Chocka-lingam, "A Fundamental Study on the Aerodynamics of Four Middle and Long Distance Running Shoes," *International Journal of Sports Science and Engineering* 5, no. 2 (2011): 119–128.

32. Helen Chandler, "Controversy over New World Cup Ball," *CNN*, June 3, 2010, http://www.cnn.com/2010/SPORT/football/06/02/football.jabulani.ball.world.cup /index.html.

33. Firoz Alam, Harun Chowdhury, Hazim Moria, Franz Konstantin Fuss, Iftekhar Khan, Fayez Aldawi, Aleksandar Subic, "Aerodynamics of Contemporary FIFA Soccer Balls," *Procedia Engineering* 13, no. 2 (2011): 188–193.

34. J. N. Gelberg, "The Rise and Fall of the Polara Asymmetric Golf Ball: No Hook, No Slice, No Dice," *Technology in Society* 18, no. 1 (1996): 93–110.

35. "NBA Introduces New Game Ball," June 28, 2006, http://www.nba.com/news /blackbox_060628.html.

36. Ibid.

37. Horween Leathers, "About," http://horween.com/about/.

38. "NBA Introduces New Game Ball."

39. "The NBA's New Basketball: What They're Saying," June 28, 2006, http://www .nba.com/news/newball_quotes_060628.html.

40. Ibid.

41. Kaushik De and Jim Horwitz, "MavBalls Investigation," October 26, 2006, http://heppc1.uta.edu/kaushik/mavstudy/summary.htm.

42. Marc Stein, "NBA Ball Controversy Reaches New Level," *ESPN*, December 8, 2006, http://sports.espn.go.com/nba/columns/story?columnist=stein_marc&id =2689744.

43. Roscoe Nance, "NBA to Ditch New Ball, Return to Old," *USA Today*, Decem-ber 11, 2006, http://usatoday30.usatoday.com/sports/basketball/nba/2006-12-11-new -ball-gone_x.htm.

44. Liz Robbins, "A Whole New Game Ball? N.B.A. Admits Its Mistake," Decem-ber 6, 2006, http://www.nytimes.com/2006/12/06/sports/basketball/06ball.html.

45. Schmidt, "N.B.A. to Take Up Complaints with Ball Manufacturer."

46. Robbins, "A Whole New Game Ball? N.B.A. Admits Its Mistake."

47. Ibid.

48. Marc Stein, "Leather Ball Will Return on Jan. 1," *ESPN*, December 12, 2006, http://sports.espn.go.com/nba/news/story?id=2694335.

49. Jay Jaffee, "What *Really* Happened in the Juiced Era," in *Extra Innings: More Baseball between the Numbers from the Team at Baseball Prospectus*, ed. Steven Goldman (New York: Basic Books, 2012), 53–97; Theodore V. Wells Jr., Brad S. Karp, and

Lorin L. Reisner, "Investigative Report concerning Footballs Used during the AFC Championship Game on January 18, 2015," https://nfllabor.files.wordpress.com/2015/05/investigative-and-expert-reports-re-footballs-used-during-afc-championsh.pdf.

50. Paul M. Sommers, Robert M. Marcoux, Filip Marinkovic, and George A. Mayer, "Why the NBA Bounced Their New Ball," Middlebury College Economics Discussion Paper No. 07-19, http://econpapers.repec.org/paper/mdlmdlpap/0719.htm.

51. "'Estoy Satisfecho de mi Hazaña' Manifestó el Belga Merckx," El Informador, October 26, 1972, 4, section B.

52. Union Cycliste Internationale, "The Lugano Charter," October 8, 1996, http://oldsite.uci.ch/imgarchive/Road/Equipment/The%20Lugano%20charter.pdf.

53. Union Cycliste Internationale, "Minimum Bicycle Weight," June 11, 1999, http://oldsite.uci.ch/english/news/news_pre2000/comm_19990611.htm.

54. Union Cycliste Internationale, "UCI Hour Record," September 9, 2000, http://oldsite.uci.ch/english/news/news_pre2000/comm_20000908.htm.

55. Tim Maloney, "Ernesto Colnago 50th Anniversary Interview: Part Four," Cyclingnews, March 19, 2004, http://autobus.cyclingnews.com/sponsors/italia/2004/colnago/?id=colnago4.

56. Owen Mulholland, "Eddy and the Hour," Bicycle Guide 8 (March 1991): 65.

57. The bicycles were still significantly lighter than the 6.8 kilogram limit instituted in January 2000 with the UCI's Lugano Charter.

58. Maloney, "Ernesto Colnago 50th Anniversary Interview: Part Four."

59. Le Monde, October 25, 1972, 16; Corriere Della Sera, October 25, 1972, p. 20; L'Aurore, October 26, 1972, p. 11A; The New York Times, October 26, 1972, p. 54.

60. "Bike with Engine (Doped Bike) and Cancellara (Roubaix - Vlaanderen)," YouTube, May 29, 2010, https://youtu.be/8Nd13ARuvVE.

61. "Cancellara Caught??" Bikeradar.com Blog, accessed October 13, 2015, http://www.bikeradar.com/forums/viewtopic.php?t=12705314.

62. Henry Samuel, "Fabian Cancellara Denies Using a Motorised Bike to Win Races in the Spring," The Telegraph, June 2, 2010, http://www.telegraph.co.uk/sport/othersports/cycling/7798626/Fabian-Cancellara-denies-using-a-motorised-bike-to-win-races-in-the-spring.html.

63. Ibid.

64. Shane Stokes and Stephen Farrand, "UCI Denies Reports of Motorised Doping," Cyclingnews, June 2, 2010, http://www.cyclingnews.com/news/uci-denies-reports-of-motorised-doping.

65. Ibid.

66. Ibid.

67. "UCI to Scan Tour Bikes for Motors," Irish Examiner, June 18, 2010, http://www.irishexaminer.com/breakingnews/sport/uci-to-scan-tour-bikes-for-motors-462261.html; "UCI is Checking the Conformity of Approved Frames and Forks in the Field," July 17, 2012, http://www.uci.ch/news/article/uci-checking-the-conformity-approved-frames-and-forks-the-field/.

68. Simon MacMichael, "Mechanical Doping—The Pro Cycling Story that Won't Go Away," Road.cc, July 2, 2015, http://road.cc/content/news/155929-mechanical-doping-pro-cycling-story-wont-go-away.

69. Jered Gruber, "Fabian Cancellara's 'So Called' Motorized Doping Debunked," *Velonation*, June 3, 2010, http://www.velonation.com/News/ID/4415/Fabian -Cancellaras-so-called-motorized-doping-debunked.aspx.

70. Ron George, "GRUBER Assist Made No Sale to Cancellara," *Cozy Beehive*, June 2, 2010, http://cozybeehive.blogspot.com/2010/06/gruber-assist-made-no-sale-to.html.

71. UCI Cycling Regulations, Part 12 Discipline and Procedures, Version 30.01.2015, accessed October 13, 2015, http://www.uci.ch/mm/Document/News/Rulesandregulation /16/26/68/12-DIS-20150101-E_English.pdf.

72. Daniel McMahon, "A 19-Year-Old Belgian Cyclist Got Caught Cheating at the World Championships after Racing a Bike That Had a Motor Hidden in the Frame," *Business Insider*, February 1, 2016, http://www.businessinsider.com/bike-investigated -technological-fraud-cycling-world-championships-cyclocross-2016-1.

73. Zachary D. Rymer, "The Evolution of the Baseball from the Dead-Ball Era through Today," *Bleacher Report*, June 18, 2013, http://bleacherreport.com/articles /1676509-the-evolution-of-the-baseball-from-the-dead-ball-era-through-today.

Chapter 3. Disabled, Superabled, or Normal?

1. Pistorius punctuated his amazing 2012 in an interesting way on December 12 by racing an Arabian horse in Doha, Qatar. This event, which he won, has an eerie resemblance to Jesse Owens performing in similar exhibitions after his 1936 Olympic triumphs. See http://youtu.be/m1L9McKLf38.

2. The only other athlete known to have competed and won Olympic medals with some form of prosthetic limb (a wooden left leg) is George Eyser, who won six gold medals in gymnastics at the 1904 Summer Olympics in St. Louis. See Amanda K. Booher, "Docile Bodies, Supercrips, and the Plays of Prosthetics," *International Journal of Feminist Approaches to Bioethics* 3, no. 2 (2010): 63–89.

3. Robert McRuer, "Compulsory Able-Bodiedness and Queer/Disabled Existence," in *The Disability Studies Reader*, ed. Lennard J. Davis (New York: Routledge, 2013), 369–380.

4. Oliver Pickup, "London 2012 Olympics: Games Legend Michael Johnson Believes Oscar Pistorius Has an 'Unfair Advantage,'" *The Telegraph*, July 17, 2012, http://www .telegraph.co.uk/sport/olympics/athletics/9405113/London-2012-Olympics-Games -legend-Michael-Johnson-believes-Oscar-Pistorius-has-an-unfair-advantage.html.

5. Abram Anders, "Foucault and 'the Right to Life': From Technologies of Normal-ization to Societies of Control," *Disability Studies Quarterly* 33, no. 3 (2013), http://dsq -sds.org/article/view/3340/3268.

6. Lindsay Berra, "Force of Habit," *ESPN The Magazine*, April 2, 2012, p. 94.

7. Ibid.

8. Paul Ducheyne, Robert L. Mauck, and Douglas H. Smith, "Biomaterials in the Repair of Sport Injuries," *Nature Materials* 11, no. 8 (2012): 652.

9. *PGA Tour, Inc. v. Martin*, 532 U.S. 661 (2001).

10. Ibid., 685.

11. Ibid., 686.

12. Michael Whitmer, "Casey Martin's Return to US Open Inspirational," *The Boston Globe*, June 12, 2012, http://www.bostonglobe.com/sports/2012/06/12/been-quite-ride -for-casey-martin/27YrPyV7YhfmZozpAeDsTK/story.html.

13. Discussions around the augmentation of bodies has moved well beyond athletics to the reconfiguration of humans in a highly technoscientific way. https://www.nextnature.net/themes/augmented-bodies/.

14. Emma M. Beckman, Mark J. Connick, Mike J. McNamee, Richard Parnell, and Sean M. Tweedy, "Should Markus Rehm be Permitted to Compete in the Long Jump at the Olympic Games?" *British Journal of Sports Medicine*, July 29, 2016, doi:10.1136/bjsports-2016-096621; Alan J. Levinovitz, "In an Era of Doping and Blade Running, What is a 'Natural' Athlete, Anyway?" *The Washington Post*, August 1, 2016, https://www.washingtonpost.com/national/health-science/in-an-era-of-doping-and-blade-running-what-is-a-natural-athlete-anyway/2016/08/01/a675e3e2-42b6-11e6-88d0-6adee48be8bc_story.html.

15. John Harris, *Enhancing Evolution: The Ethical Case for Making Better People* (Princeton, NJ: Princeton University Press, 2010): 86–108.

16. Clare Harvey, "What's Disability Got to Do with It? Changing Constructions of Oscar Pistorius before and after the Death of Reeva Steenkamp," *Disability and Society* 30, no. 2 (February 2015): 299–304.

17. Mike Burnett, "Olympic Dreams of a Blade Runner," *BBC Sport*, May 5, 2005, http://news.bbc.co.uk/sport2/hi/other_sports/disability_sport/4487443.stm.

18. P. Sewell, S. Noroozi, J. Vinney, and S. Andrews, "Developments in the Trans-tibial Prosthetic Socket Fitting Process: A Review of Past and Present Research," *Prosthetics and Orthotics International* 24, no. 2 (2000): 97–107.

19. "Bittersweet Victory: Brian Frasure on World Records, Inspiration, and Retiring on Gold," *The O and P Edge*, November 2008, http://www.oandp.com/articles/2008-11_15.asp.

20. Martha Davidson, "Innovative Live: Artificial Parts: Van Phillips," *Lemelson Center for the Study of Invention and Innovation*, March 9, 2005, http://invention.si.edu/innovative-lives-artificial-parts-van-phillips.

21. Felix Gillette, "Racing Tall: A Paralympian Stands Accused of Getting an Illegal Leg Up," *Slate*, November 10, 2004, http://www.slate.com/articles/sports/left_field/2004/11/racing_tall.html.

22. Ibid.

23. It can be argued that this stride length advantage is one reason why Usain Bolt has been one of the most dominant sprinters of all time. See Nicholas Romanov, "Analysis of Usain Bolt's Running Technique," *Pose Method*, https://posemethod.com/usain-bolts-running-technique/.

24. Ralph Beneke and Mathew J. Taylor, "What Gives Bolt the Edge—A. V. Hill Knew It Already!" *Journal of Biomechanics* 43, no. 11 (2010): 2241–2243.

25. *IPC Athletics Classification Rules and Regulations*, version 9 (2011): 46–47.

26. "IAAF Council Introduces Rule Regarding 'Technical Aids,'" *IAAF*, March 26, 2007, http://www.iaaf.org/news/news/iaaf-council-introduces-rule-regarding-techni.

27. Jeré Longman, "An Amputee Sprinter: Is He Disabled or Too-Abled?" *The New York Times*, May 15, 2007, http://www.nytimes.com/2007/05/15/sports/othersports/15runner.html?pagewanted=all&_r=0; Josh McHugh, "Blade Runner," *Wired*, March 2007, http://archive.wired.com/wired/archive/15.03/blade.html.

28. Court of Arbitration of Sport, "Arbitration CAS 2008/A/1480 Pistorius v/ IAAF," 6.

29. International Association of Athletics Federations, *Competition Rules* 2008, http://www.dis-sportschiedsgericht.de/files/regelwerke/IAAF_Competition_Rules _2008.pdf.

30. "IAAF Council Introduces Rule Regarding 'Technical Aids.'"

31. Court of Arbitration of Sport, "Arbitration CAS 2008/A/1480 Pistorius v/ IAAF," 3.

32. Ibid.

33. Chris R. Abbiss and Paul B. Laursen, "Describing and Understanding Pacing Strategies during Athletic Competition," *Sports Medicine* 38, no. 3 (2008): 239–252.

34. Court of Arbitration of Sport, "Arbitration CAS 2008/A/1480 Pistorius v/ IAAF," 3.

35. Ibid., 4.

36. Gert-Peter Brüggemann, Adamantios Arampatzis, Frank Emrich, and Wolfgang Potthast, "Biomechanics of Double Transtibial Amputee Sprinting Using Dedicated Sprinting Prostheses," *Sports Technology* 1, nos. 4–5 (2008): 220–227.

37. Court of Arbitration of Sport, "Arbitration CAS 2008/A/1480 Pistorius v/ IAAF," 9.

38. "Oscar Pistorius - Independent Scientific Study Concludes that Cheetah Prosthetics Offer Clear Mechanical Advantages," *IAAF*, January 14, 2008, http://www .iaaf.org/news/news/oscar-pistorius-independent-scientific-stud-1.

39. Rose Eveleth, "Should Oscar Pistorius's Prosthetic Legs Disqualify Him from the Olympics?" *Scientific American*, July 24, 2012, http://www.scientificamerican.com /article.cfm?id=scientists-debate-oscar-pistorius-prosthetic-legs-disqualify-him -olympics.

40. Peter G. Weyand, Matthew W. Bundle, Craig P. McGowan, Alena Grabowski, Mary Beth Brown, Rodger Kram, and Hugh Herr, "The Fastest Runner on Artificial Legs: Different Limbs, Similar Function?" *Journal of Applied Physiology* 107, no. 3 (2009): 903–911.

41. Court of Arbitration of Sport, "Arbitration CAS 2008/A/1480 Pistorius v/ IAAF," 9, 10, 12, 13.

42. Ibid., 7.

43. Ibid., 9.

44. Ibid., 14.

45. "Relay Safety Fears over Pistorius," *BBC Sports*, July 15, 2008, http://news.bbc.co .uk/sport2/hi/olympics/athletics/7508399.stm.

46. Joshua Robinson, "Amputee Sprinter's Beijing Quest Is Over," *The New York Times*, July 19, 2008, http://www.nytimes.com/2008/07/19/sports/olympics/19track .html?_r=0.

47. Jon Mulkeen, "Pistorius Gets World and Olympic Qualifier in Lignano," *Athletics Weekly*, July 19, 2011, https://web.archive.org/web/20160116134230/http://www .athleticsweekly.com/0/admin/news/pistorius-gets-world-and-olympic-qualifier-in -lignano/.

48. "Bladerunner Pistorius Included in SA's Olympic Team," *South African Sports Confederation and Olympic Committee*, July 4, 2012, http://www.sascoc.co.za/2012/07/04 /bladerunner-pistorius-included-in-sas-olympic-team/.

49. Court of Arbitration of Sport, "Arbitration CAS 2008/A/1480 Pistorius v/ IAAF," 10

50. Ibid., 11.

51. David Epstein, "New Study, For Better or Worse, Puts Pistorius' Trial in Limelight," *Sport Illustrated*, November 19, 2009, http://www.si.com/more-sports/2009/11/19/oscar-pistorius.

52. Peter G. Weyand and Matthew W. Bundle, "Point: Counterpoint: Artificial Limbs Do/Do Not Make Artificially Fast Running Speeds Possible," *Journal of Applied Physiology* 108, no. 4 (2010): 1011–1012.

53. Rodger Kram, Alena Grabowski, Craig McGowan, Mary Beth Brown, and Hugh Herr, "Counterpoint: Artificial Legs Do Not Make Artificially Fast Running Speeds Possible," *Journal of Applied Physiology* 108, no. 4 (2010): 1012–1014. See also Peter G. Weyand and Matthew W. Bundle, "Rebuttal from Weyand and Bundle," *Journal of Applied Physiology* 108, no. 4 (2010): 1014; Rodger Kram, Alena Grabowski, Craig McGowan, Mary Beth Brown, William McDermott, Matthew Beale, and Hugh Herr, "Rebuttal from Kram, Grabowski, McGowan, Brown, Mcdermott, Beale, and Herr," *Journal of Applied Physiology* 108, no. 4 (2010): 1014–1015; Alena M. Grabowski, Craig P. McGowan, William J. McDermott, Matthew T. Beale, Rodger Kram, and Hugh M. Herr, "Running-Specific Prostheses Limit Ground-Force during Sprinting," *Biology Letters* 6, no. 2 (2010): 201–204.

54. Eveleth, "Should Oscar Pistorius's Prosthetic Legs Disqualify Him from the Olympics?"

55. Paddy Allen, "Pistorius v Oliveira: Battle of the Blades," *The Guardian*, September 3, 2012, http://www.guardian.co.uk/sport/interactive/2012/sep/03/pistorius-oliveira-paralympics-2012-interactive.

56. Owen Gibson, "Paralympics 2012: Oscar Pistorius Erupts after Alan Oliveira Wins Gold," *The Guardian*, September 2, 2012, http://www.theguardian.com/sport/2012/sep/03/paralympics-2012-oscar-pistorius-alan-oliveira.

57. Ebenezer Samuel, "Oscar Pistorius' Image Takes a Hit after Disabled Track Star Accuses Fellow Paralympian of Having Unfair Advantage," *New York Daily News*, September 8, 2012, http://www.nydailynews.com/sports/more-sports/oscar-pistorius-image-takes-hit-disabled-track-star-accuses-fellow-paralympian-unfair-advantage-article-1.1154975.

58. Owen Gibson, "Oscar Pistorius Expressed Concerns About Blade Length Weeks Ago, Says IPC," *The Guardian*, September 3, 2012, http://www.guardian.co.uk/sport/2012/sep/03/oscar-pistorius-blade-length-ipc.

59. Owen Gibson, "Alan Oliveira Began Using Taller Blades Three Weeks before Paralympics," September 3, 2012, http://www.guardian.co.uk/sport/2012/sep/03/alan-oliveira-taller-blades-pistorius?intcmp=239.

60. Ross Tucker, "Oscar Pistorius: Counting Strides (as Requested) and More Thoughts," *The Science of Sport*, September 3, 2012, http://www.sportsscientists.com/2012/09/oscar-pistorius-counting-strides-as.html.

61. Gibson, "Alan Oliveira Began Using Taller Blades Three Weeks before Paralympics."

62. Ibid.

63. Ross Tucker, "Pistorius vs Oliveira and Technology. Three Quick Thoughts on Round 4," *The Science of Sport*, September 8, 2012, http://www.sportsscientists.com/2012/09/pistorius-vs-oliveira-and-technology-3.html.

64. Ibid.

65. Shawn M. Crincoli, "You Can Only Race If You Can't Win? The Curious Cases of Oscar Pistorius and Caster Semenya," *Texas Review of Entertainment and Sports Law* 133 (2011): 181.

66. Ibid., 180.

67. Patricia J. Zettler, "Is it Cheating to Use Cheetahs? The Implications of Technologically Innovative Prostheses for Sports Values and Rules," *Boston University International Law Journal* 27 (2009): 397.

68. Siobhan Sommerville, "Queer," in *Keyword for American Cultural Studies*, ed. Bruce Burgett and Glenn Hendler (New York: New York University Press, 2007), 187–191.

69. Ivo van Hilvoorde and Laurens Landeweerd, "Disability or Extraordinary Talent—Francesco Lentini (Three Legs) versus Oscar Pistorius (No Legs)," *Sport, Ethics and Philosophy* 2, no. 2 (August 2008): 99.

70. Amanda K. Booher, "Defining Pistorius," *Disability Studies Quarterly* 31, no. 3 (2011): 6.

Chapter 4. "I Know One When I See One"

1. Danielle Peers, "Patients, Athletes, Freaks: Paralympism and the Reproduction of Disability," *Journal of Sport and Social Issues* 36, no. 3 (2012): 295–316; Michael M. Chemers, "Staging Stigma: A Freak Studies Manifesto," *Disability Studies Quarterly* 25, no. 3 (2005): http://dsq-sds.org/article/view/574/751; John Harris, "The Biggest, Strongest and Most Athletic Freaks from the 2016 NFL Scouting Combine," *The Washington Post*, March 3, 2016, https://www.washingtonpost.com/news/sports/wp/2016/03/03/the-biggest -strongest-and-most-athletic-freaks-from-the-2016-nfl-scouting-combine/.

2. Tara Magdalinski, *Sport, Technology and the Body: The Nature of Performance* (New York: Routledge, 2009), 14–30.

3. Donna J. Haraway, *Simians, Cyborgs, and Women: The Reinvention of Nature* (New York: Routledge, 1991), 154.

4. Brenna Munro, "Caster Semenya: Gods and Monsters," *Safundi: The Journal of South African and American Studies* 11, no. 4 (2010): 383–396; Tavia Nyong'o, "The Unforgivable Transgression of Being Caster Semenya," *Women and Performance: A Journal of Feminist Theory* 20, no. 1 (2010): 95–100; John M. Sloop, "'This is Not Natural': Caster Semenya's Gender Threats," *Critical Studies in Media Communication* 29, no. 2 (2012): 81–96; Neville Hoad, "'Run, Caster Semenya, Run!' Nativism and the Translations of Gender Variance," *Safundi: The Journal of South African and American Studies* 11, no. 4 (2010): 397–405; Jayne Caudwell, ed., *Sport, Sexualities, and Queer Theory* (New York: Routledge, 2006).

5. Haraway, *Simians, Cyborgs, and Women*, 151.

6. Paul Dimeo, *A History of Drug Use in Sport 1876–1976: Beyond Good and Evil* (New York: Routledge, 2007), 53–86.

7. Eileen McDonagh and Laura Pappano, *Playing with the Boys: Why Separate is not Equal in Sports* (New York: Oxford University Press, 2008).

8. Kosuke Kojima, Paul L. Jamison, and Joel M. Stager, "Multi-age-grouping Paradigm for Young Swimmers," *Journal of Sports Sciences* 30, no. 3 (2012): 313–320.

9. Heidi Eng, "Issues of Gender and Sexuality in Sport," in *Gender and Sport: Changes and Challenges*, ed. Gertrud Pfister and Mari Kristin Sisjord (Münster: Waxmann, 2013): 159–173.

10. Susan Birrell and Cheryl L. Cole, "Double Fault: Renee Richards and the Construction and Naturalization of Difference," *Sociology of Sport Journal* 7, no. 1 (1990): 1–21.

11. Gail Bederman, *Manliness and Civilization: A Cultural History of Gender and Race in the United States, 1880–1917* (Chicago: University of Chicago Press, 1996).

12. Geoffrey C. Ward, *Unforgivable Blackness: The Rise and Fall of Jack Johnson* (New York: Vintage, 2006); Jack Johnson, *My Life and Battles*, trans. Christopher Rivers (Washington, DC: Potomac Books, 2009).

13. Rupert Neate and Owen Gibson, "Olympics Anti-Doping Operation to Carry Out Record Number of Tests," *The Guardian*, July 25, 2012, http://www.guardian.co.uk/sport /2012/jul/25/olympics-anti-doping-operation-tests.

14. Rebecca R. Ruiz, "Rio Drug-Testing Lab Is Suspended by Antidoping Regulator," *The New York Times*, June 24, 2016, http://www.nytimes.com/2016/06/25/sports /olympics/rio-drug-testing-lab-is-suspended-by-wada.html.

15. Ben Rumsby, "Rio 2016 Olympics: Anti-doping Branded 'Worst' in Games History," *The Telegraph*, August 17, 2016, http://www.telegraph.co.uk/olympics/2016/08 /17/rio-2016-olympics-anti-doping-branded-worst-in-games-history/.

16. Bennett Foddy and Julian Savulescu, "Time to Re-evaluate Gender Segregation in Athletics?" *British Journal of Sports Medicine* 45, no. 15 (2011): 1184–1188.

17. "Caster Semenya: Anatomy of Her Case," *The Telegraph*, July 6, 2010, http://www .telegraph.co.uk/sport/othersports/athletics/7873921/Caster-Semenya-anatomy-of-her -case.html.

18. David Smith, "Caster Semenya Row: 'Who Are White People to Question the Makeup of an African Girl? It is Racism,'" *The Guardian*, August 22, 2009, http://www .guardian.co.uk/sport/2009/aug/23/caster-semenya-athletics-gender.

19. Mike Hurst, "Caster Semenya Has Male Sex Organs and No Womb or Ovaries," *The Daily Telegraph*, September 11, 2009, http://www.dailytelegraph.com.au/sport /semenya-has-no-womb-or-ovaries/story-e6frexni-1225771672245.

20. Jacquelin Magnay, "Secret of Semenya's Sex Stripped Bare," *The Sydney Morning Herald*, September 11, 2009, http://www.smh.com.au/news/sport/secret-of-semenyas -sex-stripped-bare/2009/09/11/1252519599453.html.

21. Lucky Sindane, "Semenya Saga: Chuene's Trail of Lies," *Mail and Guardian*, September 18, 2009, http://mg.co.za/article/2009-09-18-semenya-saga-chuenes-trail -of-lies.

22. Ibid.

23. Ibid.

24. Fatima Asmal, "ASA Mentor Fights Semenya Fallout," *Mail and Guardian*, September 13, 2013, http://mg.co.za/article/2013-09-12-asa-mentor-fights-semenya -fallout.

25. Ibid.

26. Ibid.

27. Harold Adams, "Team Doctor's Report on Developments around Ms Caster Semenya re the 20009 Athletics World Championships in Berlin, Germany," http://resources.news.com.au/files/2009/10/26/1225791/515746-dt-file-caster-2.pdf.

28. Ibid.

29. Ibid.

30. Ibid.

31. Ibid.

32. "Statement on Caster Semenya," *IAAF*, September 11, 2009, http://www.iaaf.org/news/news/statement-on-caster-semenya.

33. Celean Jacobson, "SAfrica Track Chief Apologizes for Runner Sex Test," *The San Diego Union Tribune*, September 19, 2009, http://www.sandiegouniontribune.com/news/2009/sep/19/ath-semenya-gender-test-091909/.

34. Simon Hart, "IAAF Confirms Caster Semenya's Return," *The Telegraph*, July 6, 2010, http://www.telegraph.co.uk/sport/othersports/athletics/7875157/IAAF-confirms-Caster-Semenyas-return.html.

35. "Caster Semenya May Compete," *IAAF*, July 6, 2010, http://www.iaaf.org/news/iaaf-news/caster-semenya-may-compete.

36. Arne Ljungqvist and Joe Leigh Simpson, "Medical Examination for Health of All Athletes Replacing the Need for Gender Verification in International Sports," *Journal of the American Medical Association* 267, no. 6 (1992): 850–852.

37. Joe Leigh Simpson, Arne Ljungqvist, Malcolm A. Ferguson-Smith, Albert de la Chapelle, Louis J. Elsas II, Anke A. Ehrhardt, Myron Genel, Elizabeth A. Ferris, and Alison Carlson, "Gender Verification in the Olympics," *Journal of the American Medical Association* 284 (2000): 1568–1569.

38. Laura Hercher, "Gender Verification: A Term Whose Time Has Come and Gone," *Journal of Genetic Counseling* 19, no. 6 (2010): 552.

39. Ambroise Wonkam, Karen Fieggen, Raj Ramesar, "Beyond the Caster Semenya Controversy: The Case of the Use of Genetics for Gender Testing in Sport," *Journal of Genetic Counseling* 19, no. 6 (2010): 545.

40. Tian Qinjie, He Fangfang, Zhou Yuanzheng, and Ge Qinsheng, "Gender Verification in Athletes with Disorders of Sex Development," *Gynecological Endocrinology* 25, no. 2 (2009): 117–121.

41. Alice Dreger, "Sex Typing for Sport," *Hastings Center Report* 40, no. 2 (2010): 23.

42. Ross Tucker and Jonathan Dugas, "800m Caster Semenya and Robby Andrews," *The Science of Sport*, June 12, 2011, http://scienceofsport.blogspot.com/2011_06_01_archive.html.

43. Patrick Sawer and Sebastian Berger, "Gender Row over Caster Semenya Makes Athlete into a South African Cause Celebre," *The Telegraph* August 23, 2009, http://www.telegraph.co.uk/news/worldnews/africaandindianocean/southafrica/6073980/Gender-row-over-Caster-Semenya-makes-athlete-into-a-South-African-cause-celebre.html.

44. Silvia Camporesi, "Oscar Pistorius, Enhancement and Post-humans," *Journal of Medical Ethics* 34, no. 9 (2008): 639.

45. Thomas Hobbes, *Leviathan*, ed. with an introduction and notes by J. C. A. Gaskin (Oxford: Oxford University Press, 1998), 7.

46. Ray Kurzweil, *The Singularity Is Near: When Humans Transcend Biology* (New York: Viking, 2005).

47. Jan Rintala, "Sport and Technology: Human Questions in a World of Machines," *Journal of Sport and Social Issues* 19, no. 1 (1995): 62–75.

48. Vanessa Heggie, "Testing Sex and Gender in Sports; Reinventing, Reimagining and Reconstructing Histories," *Endeavor* 34, no. 4 (2010): 157–163.

49. Milton Bronner, "The Girl Who Became a Bridegroom," *The Florence Times*, August 20, 1936, 5, http://news.google.com/newspapers?nid=1842&dat=19360820&id =LQ0sAAAAIBAJ&sjid=E70EAAAAIBAJ&pg=1225,1468831.

50. Samantha M. Shapiro, "Caught in the Middle," *ESPN the Magazine*, August 6, 2012, p. 41.

51. International Olympic Committee, "IOC Regulations on Female Hyperandrogenism," http://www.olympic.org/Documents/Commissions_PDFfiles/Medical_commission /2012-06-22-IOC-Regulations-on-Female-Hyperandrogenism-eng.pdf.

52. Ibid; emphasis added.

53. Ibid; emphasis added.

54. Katrina Karkazis, Rebecca Jordan-Young, Georgiann Davis, and Silvia Camporesi, "Out of Bounds? A Critique of the New Policies on Hyperandrogenism in Elite Female Athletes," *The American Journal of Bioethics* 12, no. 7 (2012): 3–16

55. Juliet Macur, "I.O.C. Adopts Policy for Deciding Whether an Athlete Can Compete as a Woman," *The New York Times*, June 23, 2012, http://www.nytimes.com /2012/06/24/sports/olympics/ioc-adopts-policy-for-deciding-whether-athletes-can -compete-as-women.html?_r=0.

56. International Olympic Committee, "IOC Regulations on Female Hyperandrogenism."

57. Macur, "I.O.C. Adopts Policy for Deciding Whether an Athlete Can Compete as a Woman."

58. Rebecca Jordan-Young and Katrina Karkazis, "You Say You're a Woman? That Should Be Enough," *The New York Times*, June 17, 2012, http://nyti.ms/1Ap9owS.

59. Peter Sönksen, B. A. Bengtsson, J. S. Christiansen, L. Sacca, Prince Alexander De Merode, Anne-Marie Kappelgaard, Lynda Fryklund, Laurent Rivier, Philip Brown, Eryl Bassett, Mike Kenward, Ross Cuneo, Jennifer Wallace, Rob Baxter, and Christian Strasburger, "GH-2000 A Methodology for the Detection of Doping with Growth Hormone and Related Substances," Special Report to the International Olympic Committee and the European Union, A Biomed 2 Project, Contract # BMH4 CT950678, 1999. https://web.archive.org/web/20070304115926/http://www.gh2004.soton.ac.uk /GH-2000%20Final%20Report.pdf.

60. Rebecca Jordan-Young and Katrina Karkazis, "The IOC's Superwoman Complex": How Flawed Sex-Testing Discriminates," *The Guardian*, July 2, 2012, http://www.theguardian.com/commentisfree/2012/jul/02/ioc-supwerwoman-complex -flawed-sextesting-policy.

61. Ibid.

62. Tobias J. Moskowitz and L. Jon Wertheim, *Scorecasting: The Hidden Influences behind How Sports Are Played and Games Are Won* (New York: Crown Archetype, 2011).

63. Allen C. Lim, James E. Peterman, and Benjamin M. Turner, Lindsey R. Sweeney, and William C. Byrnes, "Comparison of Male and Female Road Cyclists under Identical Stage Race Conditions," *Medicine and Science in Sports and Exercise* 43, no. 5 (2011): 846–52.

64. "Baylor Basketball (W): Brittney Griner's Two-Handed Dunk vs. Georgia Tech," *YouTube*, March 24, 2012, http://youtu.be/f4w9EfAFGIw.

65. "Candace Parker Dunk against Army," *YouTube*, July 25, 2006, http://youtu.be/_A-ZNfdBiqQ.

66. Rachael Larimore, "The Difference between Saying Brittney Griner 'Plays Like a Man' and Calling Her a Man," *Slate*, April 4, 2012, http://www.slate.com/blogs/xx_factor/2012/04/04/brittney_griner_plays_like_a_man_that_doesn_t_make_it_ok_to_call_her_a_man_.html.

67. Ann Killion, "Griner Shines Bright under Spotlight of Biggest Game in Her Career," *Sports Illustrated*, April 4, 2012, http://www.si.com/more-sports/2012/04/04/brittney-griner.

68. Brittney Griner, *In My Skin: My Life on and off the Basketball Court* (New York: It Books, 2014).

69. John Brenkus, "You Ain't Seen Nothing Yet: We Are Only Scratching the Surface of What Women Can Accomplish in Sports," *ESPN The Magazine*, June 11, 2012, 63–66. This article presents a fascinating set of infographics. One in particular shows how a female boxer can create more punching force than the male boxer of equal weight.

70. "Radcliffe: Caster Semenya Rio Gold 'Won't Be Sport,'" *BBC Radio 5 Live In Short*, July 21, 2016, http://www.bbc.co.uk/programmes/p0425m52.

71. Jeré Longman, "Understanding the Controversy Over Caster Semenya," *The New York Times*, August 18, 2016, http://www.nytimes.com/2016/08/20/sports/caster-semenya-800-meters.html.

72. "Radcliffe: Caster Semenya Rio Gold 'Won't Be Sport.'"

73. Guy Wilson-Roberts, "Un Petit Tour," *Le Grimpeur*, September 26, 2012, https://web.archive.org/web/20160519015529/http://le-grimpeur.net/blog/archives/738.

74. Arthur L. Caplan, "Fairer Sex: The Ethics of Determining Gender for Athletic Eligibility: Commentary on 'Beyond the Caster Semenya Controversy: The Case of the Use of Genetics for Gender Testing in Sport,'" *Journal of Genetic Counseling* 19, no. 6 (December 2010): 550.

75. Haraway, *Simians, Cyborgs, and Women*, 150.

76. Ibid., 181.

Chapter 5. The Parable of a Cancer Jesus

1. "USADA Notice Letter against Lance Armstrong," *The Wall Street Journal*, June 12, 2012, http://online.wsj.com/article/SB10001424052702303734204577464954262704154.html, and http://d3epuodzu3wuis.cloudfront.net/2012-06-12+AFLD+Notice+Letter+REDACTED.pdf.

2. "Ferrari, del Moral and Marti Banned for Life in US Postal Case," *CyclingNews*, July 10, 2012, http://www.cyclingnews.com/news/ferrari-del-moral-and-marti-banned-for-life-in-us-postal-case/.

3. Bill Gifford, "Paging Doctor Ferrari," *Bicycling* 47 (2006): 51–59.

4. "Lance Armstrong Responds to USADA Allegation," June 13, 2012, https://web
.archive.org/web/20120613211430/http://lancearmstrong.com/news-events/lance
-armstrong-responds-to-usada-allegation.

5. Willy Voet, *Breaking the Chain: Drugs and Cycling: The True History* (London:
Yellow Jersey, 2001): 1–18.

6. Ask Vest Christiansen, "The Legacy of Festina: Patterns of Drug Use in European
Cycling since 1998," *Sport in History* 25, no. 3 (2005): 497–514.

7. Shane Stokes, "USADA Denies UCI Request to Take Control of Armstrong/USPS
Doping Proceedings," *VeloNation*, August 3, 2012, http://www.velonation.com/News/ID
/12563/USADA-denies-UCI-request-to-take-control-of-ArmstrongUSPS-doping
-proceedings.aspx; Alan McLean, Archie Tse, and Lisa Waananen, "Top Finishers of the
Tour de France Tainted by Doping," *The New York Times*, January 4, 2013, http://www
.nytimes.com/interactive/2012/08/24/sports/top-finishers-of-the-tour-de-france
-tainted-by-doping.html?_r=0.

8. "Reasoned Decision of the United States Anti-Doping Agency on Disqualification
and Ineligibility," October 10, 2012, 140–141, http://d3epuodzu3wuis.cloudfront.net
/ReasonedDecision.pdf.

9. Stephen Farrand, "McQuaid Reveals Armstrong Made Two Donations to the
UCI," *CyclingNews*, July 10, 2012, http://www.cyclingnews.com/news/mcquaid-reveals
-armstrong-made-two-donations-to-the-uci.

10. Daniel Benson, "The United States of Omerta," *CyclingNews*, August 31, 2012,
http://www.cyclingnews.com/features/the-united-states-of-omerta.

11. Tony Ortega, "Lance Armstrong in Italy: Remembering Cycling's 'Omerta,'"
May 8, 2009, http://blogs.villagevoice.com/runninscared/2009/05/lance_armstrong_8
.php.

12. Sissela Bok, *Lying: Moral Choice in Public and Private Life* (New York: Pantheon
Books, 1978).

13. Arthur T. Costigan and Leslee Grey, eds., *Demythologizing Educational Reforms:
Responses to the Political and Corporate Takeover of Education* (New York: Routledge, 2014).

14. *Lance Armstrong v Travis Tygart*, *The United States District Court for the Western
District of Texas Austin Division*, CaseNo.A-12-CA-606-S, July 9, 2012, http://graphics8
.nytimes.com/packages/pdf/sports/20120709armstrongsuit/Armstrong_order.pdf?ref
=sports.

15. "US Congressman Questions Role of USADA in Armstrong Case," *CyclingNews*,
July 13, 2012, http://www.cyclingnews.com/news/us-congressman-questions-role-of
-usada-in-armstrong-case; "Senator F. James Sensenbrenner to R. Gil Kerlikowske,
Director Office of National Drug Control," Policy July 12, 2012, http://sensenbrenner
.house.gov/news/documentsingle.aspx?DocumentID=303025.

16. Jane Aubrey and Laura Weislo, "Report: Livestrong Lobbyist Questions Fairness
of USADA Case with Congressman," *CyclingNews*, July 17, 2012, http://www.cyclingnews
.com/news/report-livestrong-lobbyist-questions-fairness-of-usada-case-with
-congressman.

17. "Statement by Senator McCain on USADA Investigation of Lance Armstrong,"
July 13, 2012, http://www.mccain.senate.gov/public/index.cfm/2012/7/post-81e65d04
-d3a5-57a6-a985-0c2ebfaa3cb0.

18. "Senator Michael Rubio to the Honorable Barbara Boxer and the Honorable Diane Feinstein," August 30, 2012, https://web.archive.org/web/20120916010444 /http://sd16.senate.ca.gov/sites/sd16.senate.ca.gov/files/USADA%20Letter%20to%20 Boxer-Feinstein.pdf.

19. Laura Weislo, "Judge Sides with USADA in Armstrong Suit," *CyclingNews*, August 20, 2012, http://www.cyclingnews.com/news/judge-sides-with-usada-in -armstrong-suit.

20. Even though Armstrong chose to avoid arbitration, Johan Bruyneel and Jose "Pepe" Marti chose arbitration. The result of their cases did not alter public opinion for or against Lance Armstrong.

21. Paul Kimmage, "Do the People Who Are Running Cycling Really Want to Clean It up?" *The Guardian*, August 25, 2012, http://www.guardian.co.uk/commentisfree/2012 /aug/26/cycling-clean-up/print.

22. Verner Møller, Ivan Waddington, and John M. Hoberman, eds., *Routledge Handbook of Drugs and Sport* (New York: Routledge, 2015).

23. Terry Todd, "Anabolic Steroids: The Gremlins of Sport," *Journal of Sport History* 14, no. 1 (1987): 87–107.

24. Werner W. Franke and Brigitte Berendonk, "Hormonal Doping and Androgen- ization of Athletes: A Secret Program of the German Democratic Republic Govern- ment," *Clinical Chemistry* 43, no. 7 (1997): 1262–1279; Mark Fainaru-Wada and Lance Williams, *Game of Shadows: Barry Bonds, BALCO, and the Steroids Scandal that Rocked Professional Sports* (New York: Gotham Books, 2006): 88–143; Richard H. McLaren, "Independent Person WADA Investigation of Sochi Allegations," July 16, 2016, https://wada-main-prod.s3.amazonaws.com/resources/files/20160718_ip_report _newfinal.pdf.

25. Timothy L. Epstein, "NFL Compromises on New Doping, Substances Policy," *Chicago Daily Law Bulletin* 160, no. 197 (October 7, 2014) http://www.salawus.com/media /publication/37_Epstein%20-%20NFL%20compromises%20on%20new%20doping,%20 substances%20policy%20-%20CDLB%20-%20Oct%207%202014.pdf; Wolfgang S. Weber, "Preserving Baseball's Integrity through Proper Drug Testing: Time for the Major League Baseball Players Association to Let Go of Its Collective Bargaining Reins," *University of Colorado Law Review*, 85, no. 1 (Winter 2014): 268–312.

26. Selena Roberts and David Epstein, "SI Reports New Information in the Case against Lance Armstrong," *Sports Illustrated*, May 23, 2011, http://www.si.com/more -sports/2011/01/18/lance-armstrong; Selena Roberts and David Epstein, "Was Armstrong Too Big to Fail?" *Sports Illustrated* May 23, 2011, https://web.archive.org/web /20110526011206/http://sportsillustrated.cnn.com/2011/magazine/05/23/lance .armstrong/index.html.

27. Tyler Hamilton and Daniel Coyle, *The Secret Race: Inside the Hidden World of the Tour de France: Doping, Cover-Ups, and Winning at All Costs* (New York: Bantam, 2012), 21.

28. Gregory Britton, "September 11, American 'Exceptionalism' and the War in Iraq," *Australasian Journal of American Studies* 25, no. 1 (July 2006): 125–141; Patrick Smith, *Time No Longer: Americans after the American Century* (New Haven, CT: Yale University Press, 2013).

29. Lance Armstrong and Sally Jenkins, *It's Not About the Bike: My Journey Back to Life* (New York: Berkley, 2000).

30. Paul Kimmage, "Cycling: Big Reveal of Cancer Jesus," *Irish Independent*, October 21, 2012, http://www.independent.ie/sport/other-sports/cycling-big-reveal-of-cancer-jesus-28822487.html.

31. Ibid.

32. Chris Cooper, *Run, Swim, Throw, Cheat: The Science behind Drugs in Sport* (New York: Oxford University Press, 2012), 265.

33. Andrew Hood, "How Many Times Was Armstrong Tested?" *VeloNews*, October 11, 2012, http://velonews.competitor.com/2012/10/news/how-many-times-was-armstrong-tested_256685.

34. Mark Daly, "How I Became a Drug Cheat Athlete to Test the System," *BBC News Scotland*, June 4, 2015, http://www.bbc.com/news/uk-scotland-32983932.

35. "WADA Statement on BBC Panorama Programme," *WADA*, June 3, 2015, https://www.wada-ama.org/en/media/news/2015-06/wada-statement-on-bbc-panorama-programme.

36. Ibid.

37. Hamilton and Coyle, *The Secret Race*, 125–136.

38. Victor Conte, "Conte's Prescription for Success," *BBC Sport*, May 16, 2008, http://news.bbc.co.uk/sport2/hi/olympics/athletics/7403158.stm.

39. Ibid.

40. Ibid.

41. Thomas M. Hunt, "Sport, Drugs, and the Cold War: The Conundrum of Olympic Doping Policy, 1970–1979," *Olympika* 16 (2007): 19–42; Rick Collins, "Changing the Game: The Congressional Response to Sports Doping via the Anabolic Steroid Control Act," *New England Law Review* 40, no. 3 (2006): 753–764.

42. Joe Lindsey, "CEO of USADA Travis Tygart," *ESPN The Magazine*, December 12, 2012, 104.

43. Sharra L. Vostral, *Under Wraps: A History of Menstrual Hygiene Technology* (New York: Lexington Books, 2008), 10–11.

44. Mark Jenkins, "Technological Discontinuities and Competitive Advantage: A Historical Perspective on Formula 1 Motor Racing 1950–2006," *Journal of Management Studies* 47, no. 5 (2010): 884–910.

45. James Halt, "Where is the Privacy in WADA's Whereabouts Rule?" *Marquette Sports Law Review* 20, no. 1 (2009): 268.

46. "Reasoned Decision of the United States Anti-Doping Agency on Disqualification and Ineligibility," 140–141.

47. Neal Rogers, "Analysis: What USADA's Case File Means to Those Involved," *VeloNews*, October 11, 2012, http://velonews.competitor.com/2012/10/news/analysis-what-usadas-case-file-means-to-those-involved_256676.

48. Ibid.

49. "Reasoned Decision of the United States Anti-Doping Agency on Disqualification and Ineligibility," 5.

50. "Nike Statement on Lance Armstrong," October 17, 2012, http://nikeinc.com/press-release/news/nike-statement-on-lance-armstrong.

51. Ibid.

52. William Fotheringham, "Lance Armstrong Case: Nike About-turn Completes the Disintegration," *The Guardian*, October 17, 2012, http://www.guardian.co.uk/sport /2012/oct/17/lance-armstrong-nike-livestrong-charity.

53. Ross Tucker, "Sponsors Overboard and a Guest Post on the Legalized Doping, the Armstrong Dilemma," *The Science of Sport*, October, 18, 2012, http://www .sportsscientists.com/2012/10/sponsors-overboard-guest-post-on.html.

54. Kurt Badenhausen, "Kobe Bryant's Sponsorship Will Rebound," *Forbes*, September 3, 2004, http://www.forbes.com/2004/09/03/cz_kb_0903kobe.html.

55. Tripp Mickle, "Knight on Tiger: Endorsements Come with Risk," *Street and Smith's Sports Business Journal*, December 14, 2009, http://www.sportsbusinessdaily .com/Journal/Issues/2009/12/20091214/This-Weeks-News/Knight-On-Tiger -Endorsements-Come-With-Risk.aspx.

56. Don Van Natta Jr., "Knight Reverses Course on Paterno," *ESPN*, February 11, 2013, http://espn.go.com/espn/otl/story/_/id/8934980/phil-knight-nike-says-joe -paterno-wronged-freeh-report.

57. Jeff Bercovici, "Nike Signs Michael Vick to New Endorsement Deal . . . Quietly," *Forbes*, July 5, 2011, http://www.forbes.com/sites/jeffbercovici/2011/07/05/nike-signs -michael-vick-to-new-endorsement-deal-quietly/.

58. Loreen N. Olson, Joy L. Daggs, Barbara L. Ellevold, and Teddy K. K. Rogers, "Entrapping the Innocent: Toward a Theory of Child Sexual Predators' Luring Communication," *Communication Theory* 17, no. 3 (August 2007): 231–251; John Levendis, "Qualities and Effective-Quantities of Slaves in New Orleans," *Southwestern Economic Review* 34, no. 1 (2007): 161–177.

59. *Paniagua* is a term used by Tyler Hamilton in discussing racing without using doping products. See Hamilton and Coyle, *The Secret Race*, 45.

60. Union Cycliste Internationale, "Decision of the UCI regarding the Case United States Anti-Doping Agency (USADA) versus Lance Armstrong," October 22, 2012, https://web.archive.org/web/20121024213408/http://www.uci.ch/Modules/BUILTIN /getObject.asp?MenuId=MTYzMDQ&ObjTypeCode=FILE&type=FILE&id =ODE5MjI&LangId=1.

61. "UCI Officially Nullifies Armstrong's Tour de France Titles and Results," *CyclingNews*, December 11, 2012, http://www.cyclingnews.com/news/uci-officially -nullifies-armstrongs-tour-de-france-titles-and-results.

62. Tom English, "Interview: Frozen Out of Road Cycling for Not Doping, Graeme Obree Is Vindicated and Ready for a New Challenge," *The Scotsman*, November 4, 2012, http://www.scotsman.com/scotland-on-sunday/sport/other-sports/interview-frozen -out-of-road-cycling-for-not-doping-graeme-obree-is-vindicated-and-ready-for-a-new -challenge-1-2613675.

63. Ibid.

Chapter 6. *"May I See Your Passport?"*

1. Wolfgang S. Weber, "Preserving Baseball's Integrity through Proper Drug Testing: Time for the Major League Baseball Players Association to Let Go of Its

Collective Bargaining Reins," *University of Colorado Law Review* 85, no. 1 (Winter 2014): 267–312.

2. Frank Dunne, "The Drug Scandal that Blackens the Name of Juve's Team of the Nineties," *The Independent*, December 1, 2004, http://www.independent.co.uk/news /world/europe/the-drug-scandal-that-blackens-the-name-of-juves-team-of-the -nineties-728710.html.

3. "List of Doping Cases in Cycling," http://www.self.gutenberg.org/articles/list_of _doping_cases_in_cycling.

4. From 1965 to 1992, the UCI had two main suborganizations: the Fédération Internationale Amateur de Cyclisme, for amateur cyclists, and the Fédération Internationale de Cyclisme Professional, for professionals.

5. Pierre-Edouard Sottas, Neil Robinson, Oliver Rabin, and Martial Saugy, "The Athlete Biological Passport," *Clinical Chemistry* 57, no. 7 (July 2011): 969–976.

6. "The Athlete Biological Passport," *Union Cycliste Internationale*, July 24, 2014, http://www.uci.ch/clean-sport/the-athlete-biological-passport-abp/.

7. Pierre-Edouard Sottas, Neil Robinson, Giuseppe Fischetto, Gabriel Dollé, Juan Manuel Alonso, and Martial Saugy, "Prevalence of Blood Doping in Samples Collected from Elite Track and Field Athletes," *Clinical Chemistry* 57, no. 5 (May 2011): 762–769.

8. *The IAAF Anti-Doping Athletes' Guide*, January 2015, http://www.iaaf.org /download/download?filename=c20e7184-4c7b-4433-8f83-264c0c157d44.pdf&urlslug =The%20IAAF%20Anti-Doping%20Athlete's%20Guide.

9. "Dvorak, Saugy Outline Anti-Doping Strategies," June 5, 2014, http://www.fifa .com/worldcup/news/y=2014/m=6/news=dvorak-saugy-outline-anti-doping-strategies -2354961.html.

10. Nicholas Hailey, "A False Start in the Race against Doping in Sport: Concerns with Cycling's Biological Passport," *Duke Law Journal* 61, no. 2 (2011): 393–432.

11. "Prohibited List," https://www.wada-ama.org/en/what-we-do/prohibited-list.

12. James A. R. Nafziger, "Circumstantial Evidence of Doping: BALCO and Beyond," *Marquette Sports Law Review* 16, no. 1 (Fall 2005): 51.

13. "Article 2 Anti-Doping Rule Violation," https://wada-main-prod.s3.amazonaws .com/resources/files/LEGAL_code_appendix.pdf.

14. Sean Ingle, "How Cheats Cheat: Why Dopers Have the Edge in Athletics' War on Drugs," *The Guardian*, August 20, 2015, http://www.theguardian.com/sport/2015/aug/20 /doping-world-athletics-championships-cheats.

15. David E. Newton, *Steroids and Doping in Sports: A Reference Handbook* (Santa Barbara, CA: ABC-CLIO, 2013).

16. Hein F. M. Lodewijkx and Bram Brouwer, "Some Empirical Notes on the Epo Epidemic in Professional Cycling," *Research Quarterly for Exercise and Sport* 82, no. 4 (December 2011): 740–754.

17. Charles E. Yesalis, *Anabolic Steroids in Sport and Exercise* (Champaign, IL: Human Kinetics Publishers, 1993).

18. Francesco Conconi, *Moser's Hour Record: A Human and Scientific Adventure* (Brattleboro, VT: Vitesse Press, 1991), 2.

19. I. Casoni, G. Ricci, E. Ballarin, C. Borsetto, G. Grazzi, C. Guglielmini, F. Manfredini, G. Mazzoni, M. Patracchini, E. De Paoli Vitali, F. Rigolin, S. Bartalotta, G. P. Franzè, M. Masotti, and F. Conconi, "Hematological Indices of Erythropoietin Administration in Athletes," *International Journal of Sports Medicine* 14, no. 6 (1993): 307–311.

20. Michael J. Joyner, "VO$_2$MAX, Blood Doping, and Erythropoietin," *British Journal of Sports Medicine* 37, no. 3 (June 2003): 190–191.

21. Bill Gifford, "Paging Doctor Ferrari," *Bicycling* (January/February 2006): 52, http://www.billgifford.com/wp-content/uploads/2015/03/ferrarifinal.pdf.

22. Stephen Ferrand, "Michele Ferrari Facing Trial for Doping in Italy," *Cyclingnews*, November 19, 2015, http://www.cyclingnews.com/news/michele-ferrari-facing-trial-for -doping-in-italy/; "Armstrong Defends Ferrari Friendship," *VeloNews*, June 13, 2012, http://velonews.competitor.com/2011/09/news/armstrong-defends-ferrari-friendship _193510.

23. Lawrence M. Fischer, "Stamina-Building Drug Linked to Athletes' Deaths," *The New York Times*, May 19, 1991, http://www.nytimes.com/1991/05/19/us/stamina -building-drug-linked-to-athletes-deaths.html.

24. N. Robinson, S. Giraud, C. Saudan, N. Baume, L. Avois, P. Mangin, and M. Saugy, "Erythropoietin and Blood Doping," *British Journal of Sports Medicine* 40, supplement 1 (July 2006): i30-i34.

25. Heiko Stark and Stefan Schuster, "Comparison of Various Approaches to Calculating the Optimal Hematocrit in Vertebrates," *Journal of Applied Physiology* 113, no. 3 (2012): 355–367.

26. Willy Voet, *Breaking the Chain: Drugs and Cycling: The True History* (London: Yellow Jersey, 2001), 91.

27. Jakob Mørkeberg, "Nordic Winter Sports, Biological Passport (ABP) and the Legal Challenges," https://www.uni-freiburg.de/universitaet/einzelgutachten /symposium_freiburg_2011_morkeberg.pdf; Mario Zorzoli, "Biological Passport Parameters," *Journal of Human Sport and Exercise* 6, no. 2 (2011): 205–217.

28. Charlotte Smith, "Tour du Dopage: Confessions of Doping Professional Cyclists in a Modern Work Environment," *International Review for the Sociology of Sport*, March 2, 2015, doi:10.1177/1012690215572855.

29. Voet, *Breaking the Chain*, 1–18.

30. "The Drug Scandal Widens," *Cyclingnews*, July 24, 1998, http://autobus .cyclingnews.com/results/1998/jul98/jul24.shtml; Alexandre Prévot, "Dutch Cyclists Admit Buying EPO for Infamous 1998 Tour de France," *The Amsterdam Herald*, July 11, 2014, http://www.amsterdamherald.com/index.php/rss/1212-20140711-dutch-cyclists -admit-buying-epo-1998-tour-de-france.

31. "Rapport: Au Nom de la Commission d'Enquête sur l'Efficacité de la Lutte Contre le Dopage," *Les Rapports du Sénat Français*, July 17, 2013, http://www.senat.fr/rap /r12-782-1/r12-782-11.pdf.

32. Juliet Macur, "Looking for Doping Evidence, Italian Police Raid Austrians," *The New York Times*, February 19, 2006, http://www.nytimes.com/2006/02/19/sports /olympics/19drug.html.

33. James Stray-Gundersen, Tapio Videman, Ilkka Penttilä, and Inggard Lereim, "Abnormal Hematologic Profiles in Elite Cross-country Skiers: Blood Doping or?" *Clinical Journal of Sports Medicine* 13, no. 3 (May 2003): 132–137.

34. Simon Leigh-Smith, "Blood Boosting," *British Journal of Sports Medicine* 38 (2004): 99.

35. Paul Howard, *Sex, Lies and Handlebar Tape: The Remarkable Life of Jacques Anquetil, the First Five-Times Winner of the Tour de France* (Edinburgh: Mainstream, 2011).

36. Jacques Anquetil, "The Glorious Dead," *Rouleur* 2 (2006): 82.

37. A. J. Craig Jr., "Olympics 1968: A Post-mortem," *Medicine and Science in Sport and Exercise* 1 (1969): 177–180.

38. Björn T. Ekblom, Alberto N. Goldbarg, and Bengt Gullbring, "Response to Exercise after Blood Loss and Reinfusion," *Journal of Applied Physiology* 33, no. 2 (August 1972): 180.

39. Björn T. Ekblom, "Blood Boosting and Sport," *Baillière's Clinical Endocrinology and Metabolism* 14, no. 1 (March 2000): 92.

40. Bjarne Rostaing and Robert Sullivan, "Triumphs Tainted with Blood," *Sports Illustrated*, January 21, 1985, http://www.si.com/vault/1985/01/21/546256/triumphs-tainted-with-blood.

41. Leigh-Smith, "Blood Boosting."

42. "Lausanne Declaration on Doping in Sport," *Council of Europe Report Group on Education Culture and Sport*, February 8, 1999, https://wcd.coe.int/ViewDoc.jsp?id=402791.

43. Mario Cazzola, "A Global Strategy for Prevention and Detection of Blood Doping with Erythropoietin and Related Drugs," *Haematologica* 85, no. 6 (2001): 561–563.

44. Robin Parisotto, Christopher J. Gore, Kerry R. Emslie, Michael J. Ashenden, Carlo Brugnara, Chris Howe, David T. Martin, Graham J. Trout, and Allan G, Hahn, "A Novel Method Utilizing Markers of Altered Erythropoiesis for the Detection of Recombinant Human Erythropoietin Abuse in Athletes," *Haematologica* 85, no. 6 (2001): 564–572.

45. Bradley J. Schmalzer, "A Vicious Cycle: The Biological Passport Dilemma," *University of Pittsburg Law Review* 70, no. 4 (2009): 690.

46. Mario Zorzoli and Francesca Rossi, "Implementation of the Biological Passport: The Experience of the International Cycling Union," *Drug Testing and Analysis* 2, nos. 11–12 (2010): 543.

47. Pierre-Edouard Sottas, Neil Robinson, Sylvain Giraud, Franco Taroni, Matthias Kamber, Patrice Mangin, and Martial Saugy, "Statistical Classification of Abnormal Blood Profiles in Athletes," *The International Journal of Biostatistics* 2, no. 1 (2006): 1.

48. Zorzoli and Rossi, "Implementation of the Biological Passport," 543.

49. Charles Pelkey, "Riccardo Riccò Tests Positive; Saunier Duval Team Withdraws from Tour de France," *VeloNews*, July 17, 2008, http://velonews.competitor.com/2008/07/news/road/riccardo-ricco-tests-positive-saunier-duval-team-withdraws-from-tour-de-france_80269.

50. "Center for Drug Evaluation and Research, Approval Letter," November 14, 2007, http://www.accessdata.fda.gov/drugsatfda_docs/nda/2007/125164s000_Approv .pdf; complete approval package can be found here: http://www.accessdata.fda.gov /drugsatfda_docs/nda/2007/125164toc.cfm.

51. Olivier Rabin, "Involvement of the Health Industry in the Fight against Doping in Sport," *Forensic Science International* 213, nos. 1–3 (December 2011): 10–14.

52. Séverine Lamon, Sylvain Giraud, Léonie Egli, Jessica Smolander, Michael Jarsch, Kay-Gunnar Stubenrauch, Alice Hellwig, Martial Saugy, and Neil Robinson, "A High-throughput Test to Detect C.E.R.A. Doping in Blood," *Journal of Pharmaceutical and Biomedical Analysis* 50, no. 5 (2009): 954–958.

53. "Remarks by Mr. David Howman, WADA Director General, IFPMA-WADA Joint Declaration on Cooperation in the Fight against Doping in Sport," July 6, 2010, https://wada-main-prod.s3.amazonaws.com/resources/files/remarks_dhowman _ifpma-wada_joint_declaration_06july2010.pdf.

54. "Remarks by Mr. Haruo Naito, President of the IFPMA, President and CEO of Eisai, Co., Ltd., WADA Joint Declaration on Cooperation in the Fight against Doping in Sport," July 6, 2010, https://wada-main-prod.s3.amazonaws.com/resources/files /remarks_naito_ifpma-wada_joint_declaration_06july2010.pdf.

55. "BIO Endorses 'Declaration against Doping in Sport,'" June 28, 2011, https:// www.bio.org/media/press-release/bio-endorses-"declaration-against-doping-sport.

56. "Anti-doping Collaboration Launched with 2 Fields 1 Goal Campaign," July 23, 2012, https://www.wada-ama.org/en/media/news/2012-07/anti-doping-collaboration -launched-with-2-fields-1-goal-campaign.

57. "WADA Campaign to Identify Pipeline Medicines with Doping Potential," *CyclingNews*, July 24, 2012, http://www.cyclingnews.com/news/wada-campaign-to -identify-pipeline-medicines-with-doping-potential.

58. Sottas, Robinson, Rabin, and Saugy, "The Athlete Biological Passport."

59. Carsten Lundby, Paul Robach, and Bengt Saltin, "The Evolving Science of Detection of 'Blood Doping,'" *British Journal of Pharmacology* 165, no. 5 (2012): 1306–1315.

60. John Wilcockson, "The New Passport: A Conversation with Anne Gripper," *VeloNews*, October 24, 2007, http://velonews.competitor.com/2007/10/news/the-new -passport-a-conversation-with-anne-gripper_13563.

61. Ibid.

62. Yorck Olaf Schmacher and Giuseppe d'Onofrio, "Scientific Expertise and the Athlete's Biological Passport: 3 Years of Experience," *Clinical Chemistry* 58, no. 6 (June 2012): 979–985.

63. "Information on the Biological Passport," *Union Cycliste Internationale*, December 21, 2007, https://web.archive.org/web/20110613182036/http://www.uci.ch/Modules /ENews/ENewsDetails.asp?MenuId=&id=NTQzOA.

64. Andrew Hood, "The Landis Case: Savior 'B' Samples a Rarity," *VeloNews*, August 3, 2006, http://velonews.competitor.com/2006/08/news/the-landis-case-savior -b-samples-a-rarity_10632.

65. Sottas, Robinson, Rabin, and Saugy, "The Athlete Biological Passport," 3.

66. Richard H. McLaren, "An Overview of Non-Analytical Positive and Circumstantial Evidence Cases in Sports," *Marquette Sports Law Review* 16, no. 2 (2007): 193–212.

67. Cameron A. Myler, "Resolution of Doping Disputes in Olympic Sport: Challenges Presented by 'Non-Analytical' Cases," *New England Law Review* 40, no. 3 (2006): 747–752.

68. Alex Hutchinson, "Using Math to Catch Athletes Who Dope," *The New Yorker*, August 25, 2015, http://www.newyorker.com/news/sporting-scene/using-math-to -catch-athletes-who-dope.

69. Thomas Christian Bonne, Carsten Lundby, Anne Kristine Lundby, Mikael Sander, Jacob Bejder, Nikolai Baastrup Nordsborg, "Altitude Training Causes Haematological Fluctuations with Relevance for the Athlete Biological Passport," *Drug Testing and Analysis* 7, no. 8 (August 2015): 655–662.

70. Langdon Winner, "Do Artifacts Have Politics?" *Daedalus* 109, no. 1 (1980): 121–136.

71. Carwyn Jones, "Doping in Cycling: Realism, Antirealism, and Ethical Deliberation," *Journal of the Philosophy of Sport* 37, no. 1 (2010): 88–101.

72. "Executive Committee Defines WADA Key Priorities," *World Anti-Doping Agency*, September 23, 2003, https://www.wada-ama.org/en/media/news/2003-09/executive -committee-defines-wada-key-priorities.

Conclusion. *Body, Motor, Machine*

1. Stephen G. Miller, *Ancient Greek Athletics* (New Haven, CT: Yale University Press, 2004).

2. Geoffery C. Ward, *Unforgiveable Blackness: The Rise and Fall of Jack Johnson* (New York: A. A. Knopf, 2004).

3. Robert E. Rinehart, *Players All: Performances in Contemporary Sport* (Bloomington: Indiana University Press, 1998), 1–20.

4. Verner Møller, *The Ethics of Doping and Anti-Doping: Redeeming the Soul of Sport?* (New York: Routledge, 2010); Patrick Trabal, "Resistance to Technological Innovation in Elite Sport," *International Review for the Sociology of Sport* 43, no. 3 (2008): 313–330.

5. Paul Dimeo, *A History of Drug Use in Sport 1876–1976: Beyond Good and Evil* (New York: Routledge, 2007): 33–50.

6. "Restoring Faith in America's Pastime: Evaluating Major League Baseball's Efforts to Eradicate Steroid Use," *Hearing before the Committee on Government Reform*, March 17, 2005, http://www.gpo.gov/fdsys/pkg/CHRG-109hhrg23038/html/CHRG -109hhrg23038.htm; "The Mitchell Report: The Illegal Use of Steroids in Major League Baseball," *Hearing before the Committee on Oversight and Government Reform*, January 15, 2008, http://www.gpo.gov/fdsys/pkg/CHRG-110hhrg55749/html/CHRG-110hhrg55749 .htm; "The Mitchell Report: The Illegal Use of Steroids in Major League Baseball, Day 2," *Hearing before the Committee on Oversight and Government Reform*, February 13, 2008, http://www.gpo.gov/fdsys/pkg/CHRG-110hhrg43333/html/CHRG-110hhrg43333.htm; Paul Hagen, "Arbitrator: A-Rod Suspended for 2014 Season," *MLB.com*, January 11, 2014, http://m.mlb.com/news/article/66433260/arbitrator-rules-alex-rodriguez-to-be -suspended-for-2014-season.

7. Louise Burke, Ben Desbrow, Lawrence Spriet, *Caffeine for Sports Performance: The Truth and Myths about the World's Most Popular Supplement* (Champaign, IL: Human Kinetics 2013), 109–124.

8. Juan Del Coso, Gloria Muñoz, and Jesús Muñoz-Guerra, "Prevalence of Caffeine Use in Elite Athletes Following Its Removal from the World Anti-Doping Agency List of Banned Substances," *Applied Physiology, Nutrition, and Metabolism* 36, no. 4 (2011): 555–561.

9. Lindsay Berra, "Force of Habit," *ESPN The Magazine*, April 2, 2012, 92–100.

10. "Adrian Peterson Flattered by Rumors," *ESPN*, August 13, 2013, http://espn.go .com/nfl/trainingcamp13/story/_/id/9562468/adrian-peterson-minnesota-vikings -takes-ped-rumors-compliment.

11. S. Seth Bordner, "Call 'Em as They Are: What's Wrong with Blown Calls and What to Do about Them," *Journal of the Philosophy of Sport* 42, no. 1 (2015): 101–120.

12. S. Christopher Szczerban, "Tackling Instant Replay: A Proposal to Protect the Competitive Judgments of Sports Officials," *Virginia Sports and Entertainment Law Journal* 6 (2006–2007): 277–332.

13. Steve David, "FIFA You Are Killin' Us on This Goal Line Technology Stance," *NBC Sports* March 10, 2012, http://prosoccertalk.nbcsports.com/2012/03/10/fifa-you -are-killin-us-on-this-goal-line-technology-stance/.

14. Harry Collins and Robert Evans, "You Cannot Be Serious! Public Understanding of Technology with Special Reference to 'Hawk-Eye,'" *Public Understanding of Science* 17, no. 3 (2008): 283–308.

15. Ross Tucker, "Bolt's False Start and Blake's 'Twitch'—The Actual Start Block Data Explained," *The Science of Sport*, August 30, 2011, http://www.sportsscientists.com /2011/08/bolts-false-start-and-blakes-twitch.html.

16. International Association of Athletics Federations, *IAAF Competition Rules* 2014–15, http://www.iaaf.org/about-iaaf/documents/rules-regulations, 156.

17. Ibid., 157.

18. M. Courtney, "How IT Has Become F1's Extra Gear," *Engineering and Technology* 6, no. 10 (2011): 86–89. Formula 1 is so driven by technology that its governing body has its own technical webpage (http://www.formula1.com/content/fom-website/en /championship/inside-f1/rules-regs.html) about the continuous updating of the rules or "formula." This is partially in response to the multiple journalistic outlets that follow every technical update in Formula 1.

19. Fred O. Bryant, Lee N. Burkett, Stanley S. Chen, Gary S. Krahenbuhl, and Ping Lu, "Dynamic and Performance Characteristics of Baseball Bats," *Research Quarterly: American Alliance for Health, Physical Education and Recreation* 48, no. 3 (October 1997): 505–509.

20. "Division I Baseball Statistic Trends (1970–2013)," *NCAA*, accessed October 15, 2015, http://fs.ncaa.org/Docs/stats/baseball_RB/reports/TrendsYBY.pdf.

21. Patrick Drane James Sherwood, "Baseball Studies: Baseball Bat Testing Protocol Development," *Procedia Engineering* 2, no. 2 (2010): 2681–2686.

22. Matthew Broe, James Sherwood, and Patrick Drane, "Experimental Study of the Evolution of Composite Baseball Bat Performance," *Procedia Engineering* 2, no. 2 (2010): 2653–2658.

23. NCAA, "Composite Bat Moratorium," July 17, 2009, http://www.acs.psu.edu /drussell/bats/ncaa/NCAA-Moratorium-CompositeBats-July17-2009.pdf.

24. NCAA, "Composite Bat Moratorium Is Proposed by NCAA Baseball Rules Committee," July 22, 2009, http://fs.ncaa.org/Docs/PressArchive/2009 /Announcements/20090722%2BComposite%2BBat%2BRls.html.

25. Ibid.

26. Ibid.

27. Tyler Tassone, "Should Metal Baseball Bats Come with a Warning Label? Assessing Failure to Warn Claims before and after Enactment of the BBCOR Baseball Bat Performance Standard," *Sports Lawyers Journal* 20 (Spring 2013): 211–236.

28. Jeff Bradley, "The New Deadball Era," *Sports Illustrated*, June 23, 2014, pp. 12–13.

29. The NCAA did realize this problem but troublingly introduced another technoscientific fix, the flat seam baseball, in the 2015 Division I tournament. This baseball has less aerodynamic drag, allowing it to fly farther, and hopefully will increase home runs per game. See Greg Johnson, "DI Committee Changes to Flat-seamed Baseballs for 2015 Championship," January 10, 2014, http://www.ncaa.com /news/baseball/article/2013-11-05/di-committee-changes-flat-seamed-baseballs-2015 -championship.

Index

Page numbers in *italics* refer to photographs and figures.

Adams, Harold, 137, 138, 139

Adidas: basketball shoes, 77; fast suits, 39–40, 41, 42, 46, 54, 62; formation of, 69; Jabulani soccer ball complaints, 79

Allocchio, Stefano, 95

altitude training, 190, 202

amphetamines, 133, 160, 179, 186, 211

anabolic agents, 24, 133, 160, 201. *See also* performance-enhancing substances

androgens, 141–42, 146–47, 148

Anquetil, Jacques, 190

Arena: approval under revised swimsuit regulations, 62; fast suit engineering, 40, 46, 59; Italian team sponsorship, 52, 57; open letter to FINA, 53–54, 55; Powerskin swimsuit models, 41, 46–47, 52–53; X-Glide swimsuit development, 19, 59, 63

Armstrong, Lance: admission of drug use, 155, 161, 210, 221; cancer survival, 158, 165; claims of passing drug tests, 155, 161, 162, 166, 168; Michele Ferrari relationship, 172, 185–86; heroic narrative of, 163–65, *164*, 174; jurisdiction questions in legal case, 156–57; loss of sponsorships, 173–75; public perception of, 157–58, 169, 171, 173, 174, 200; team doping leadership, 154, 172–73, 175; USADA investigation, 25–26, 154–55, 156–160, *161*, 171–73, 175, *176, 177*

athlete biological passports (ABPs), 178–204; development of, 18, 26, 176–77, 180–81, 189–191; direct testing limitations, 180, 182–83, 199–200; EPO role in need for, 184–89; evolving demands on, 197–98; future role of, 201–4; industry-wide collaborative actions, 195–97; questions about efficacy, 183–84, 202, 203–4; UCI involvement, 178–79, 180, 181–82; WADA involvement, 179, 180, 182, 191–95

Athletic Propulsion Labs (APL), 20, 74–77, 98

augmentation and assistive technology: balance among repair, assistance, and augmentation, 104–6, 124–28, 212; injury repair, 104–5, 126, 211–12; Casey Martin case, 105; Markus Rehm case, 106. *See also* Pistorius, Oscar

authenticity-body relationship, 207–18; external, maintains authenticity, 212–15; external, undermines authenticity, 215–17; graphic representation of, 208–9, *208*; internal, maintains authenticity, 211–12; internal, undermines authenticity, 209–11; overview, 17, 207

automobile racing, 5, 67, 170, 215–16, 252n18

balls, 78–86; flat seam baseballs, 253n29; golf balls, 79–80; juiced baseballs, 78–79, 86, 98; overview, 78–79; soccer balls, 79; synthetic vs. leather basketballs, 20, 80–86, 98

banned substances list: caffeine removed from, 203, 211; creation of, 192; EPO added to, 186; evolving nature of, 211; incompleteness of, 182, 198, 203

banning of technoscientific developments: basketball shoes, 20, 72–77, 98; complexity of, 12, 17–18, 98; composite bats, 218; concept of normal and, 103; cycling regulations, 20–21, 88–89, 91, 96, 98; driver aids, 67; external technoscience that undermines authenticity, 215; polyurethane swimsuits, 19, 64; synthetic basketballs, 20, 85, 98

Bannister, Roger, 5, 223n2

baseball: admissions of drug use, 178, 210–11; difficulties in historical comparisons of performance, 16–17; flat seam baseballs, 253n29; juiced baseballs, 78–79, 86, 98; testing for performance-enhancing substances, 25; ulnar collateral ligament reconstruction prevalence, 104

baseball bats, 217–220

basketball, 20, 80–86, 98, 150

basketball shoes, 20, 72–78, 98
Beard, Amanda, 51, *51*
Benjanuvatra, Nat, 44
bicycles: electric assist concerns, 93–97; hour records influenced by, 20, 87, 89–91; sublimation to bodies, 20–21; UCI revised regulations, 20–21, 88–89, 91, 96, 98. *See also* cycling
Bissinger, Buzz, 163, 165
Blake, Yohan, 213, 214
blood boosting: hematocrit level testing, 186–87, 189; lack of testing protocols, 162; rise of, 3, 184–85; through transfusions, 26, 189–191, 194, 198, 199. *See also* EPO (erythropoietin)
Bock, William, 156
body narrative: historical importance of, 3–4, 15; illusion of, 1–6, 67–68, 220; mismatches, 11; need for integrative approach with technoscience, 216–17, 222; sport authenticity questions, 207. *See also* authenticity-body relationship; human body; motor-over-machine dialectic
Bognetti, Marco, 95
Bolt, Usain, 16, 17, 109, 151, 213, 235n23
Brown, Mary Beth, 115, 121
Brüeggemann, Gert-Peter, 22, 112–13, 116, 117, 120
Brundage, Avery, 146, 149
Bruyneel, Johan, 154, 244n20
Bryant, Howard, 8
Bryant, Kobe, 174, 175
Bundle, Matthew, 115, 120, 121
business of sports: basketball shoes, 72–74, 75–76, 77; consumer vs. athlete products, 11; fast suits, 42, 43, 50–51, 51; heroic narrative role in, 205–6; performance-enhancing substances role in, 210; swimwear industry, 34, 49

caffeine, 203, 211
Cancellara, Fabian, 93–95, 96
carbon fiber materials: baseball bats, 218–19; bicycle frames, 95–96; prostheses, 21, *101*, 108, 121
Carpani, Enrico, 94
Cassani, Davide, 93–94
Castagnetti, Alberto, 52, 53
Cazzola, Mario, 193
CERA/Mircera, 195–98

cheating: public perception of, 2–3, 11, 96; testing of athletes, 134, 141, 180. *See also* performance-enhancing substances; unfair advantage
Chuene, Leonard, 137, 138, 139, 140
Clarke, Kenneth, 191
Colnago, Ernesto, 21, 89–91
competitive advantage, 10–11, 152, 216, 221. *See also* performance-enhancing substances; *specific artifacts*
Conconi, Francesco, 185, 186
Conte, Victor, 160, 167–68, 176
cotton materials, 32, 34, 35
Coughlin, Natalie, 51, *51*, 57
Court of Arbitration for Sport (CAS): appeal for Tracer B8 swimsuit approval, 63–64; Lausanne Declaration authority, 193; Oscar Pistorius case, 22, 112, 114–18, 119–120, 124, 126; Brent Rushall complaint to, 42
Crincoli, Shawn, 124–25
Cuban, Mark, 83
cultural aspects of sport: framing of, 14–18; meaning of drug testing, 162; sport meritocracy and, 6; technoscientific challenges to, 3–4, 13–14. *See also* history and tradition of sport
cyborg athletes: role of prostheses, 114, 126; sex and gender identity questions, 132, 133, 144–45, 152; in swimming, 46, 48
cycling, 86–98: Fabian Cancellara performance, 93–95, 96; cultural narratives of drug testing, 162; electric assist concerns, 93–97; EPO use, 154, 184–89, 195; European dominance of, 163; markers of excellence in, 86–87; mechanical doping term, 95, 97–98; Eddy Merckx performance, 20, 21, 87–88, 89–93, 163; motor-over-machine dialectic, 1–2; omertà tradition, 157; systematic drug use in, 155, 160, 175, 176, 178–79, 187–89, 190, 191. *See also* Armstrong, Lance; bicycles; Union Cycliste Internationale

Daly, Mark, 166–67
Dassler, Adolf ("Adi"), 20, 69–70, *70*, 72
Dassler, Rudolf, 69, 230n5
Davies, Nick, 118, 137
decision aids, 5, 212–15
Diack, Lamine, 111

direct testing for performance-enhancing substances: ABP approach vs., 199–201; banned substances list, 182, 186, 198, 203; cultural narratives of, 162; EPO testing, 186–87, 196; false positives rate, 166, 168; inefficacy of, in cycling, 155, 158, 161, 176; limitations of, 165–66, 179, 180, 182–83; methods of circumventing, 167–69, 171; move away from, 25–26, 173, 176–77; multiple purposes of, 169–171; out-of-competition testing, 176; as truth making, 177, 180
disabled, as term, 127
disabled athletes: comparisons with able-bodied athletes, 125–26, 144; dual construction with superabled athletes, 127; normative assumptions, 113; segregation of able and less-abled bodies, 100. *See also* Pistorius, Oscar
doping. *See* performance-enhancing substances
driver aids, 67
drug testing. *See* direct testing for performance-enhancing substances

EPO (erythropoietin): athlete biological passports in response to, 26, 184–89; CERA/Mircera variant of, 195–98; in cycling, 154, 184–89, 195; deaths from, 186; lack of testing protocols, 162; microdosing, 166–67, 198, 199; in soccer, 178
Eyser, George, 234n2

fair competition: dominance in sporting narratives, 12, 14; drug testing role, 169, 170, 171; evaluating bodies for, 23; governing organizations role, 169–170; misguided views of non-normative bodies, 151–52; vocal nature of publics, 10. *See also* unfair advantage
false positives, 166, 168
false starts, 213–14
Fastskin suit (Speedo), 41, 42, 44–46, 47–48, 224n31
fast suits, 37–43. *See also* full-body compression swimsuits; *specific models*
Fédération Internationale de Football Association (FIFA), 5, 79, 151, 181, 212–13
Fédération Internationale de l'Automobile (FIA), 67

Fédération Internationale de Natation (FINA): ABPs implementation, 181; fast suit concerns, 31, 32, 42, 43, 53–55, 61–62; polyurethane swimsuit ban, 19, 64; return to primacy of the body, 32, 64, 66; Speedo suit approval, 42, 50, 56, 58; swimsuit regulation revisions, 55–56, 59–66
Fédération Internationale de Ski, 187
Federazione Ciclistica Italiana, 186
Federazione Italiana Nuoto, 57
Ferrari, Michele, 154, 172, 185–86
Festina affair, 155, 187–88
Fletcher, Jennie, 33–34, 34
Forester, Pete, 73
Frasure, Brian, 108, 122
Freeman, Cathy, 102–3
full-body compression swimsuits: engineering of, 40–41, 46–48; FINA approval of, 42, 50; introduction of, 19, 39–40; revised regulations, 60, 64; scientific studies on, 44–46. *See also* LZR swimsuits

Gailey, Robert, 116–17
gender and sex verification. *See* sex and gender verification
goal line technology, 5, 213
golf, 67, 79–80, 105
Gore, Christopher J., 171–72, 173
governing organizations: anti-doping networks, 24, 210; as constituency, 9–12; role in fair competition, 10, 169–170; tension between equipment and body narrative, 15–16, 22, 67–68, 128. *See also specific organizations; specific sports*
Grabowski, Alena, 115, 121
Griffiths, Mervyn, 71
Griner, Brittney, 150
Gripper, Anne, 198–99
Gruber Assist, 95–96

Hajós, Alfréd, 33, 36
Hamilton, Tyler, 157, 163, 172, 246n59
Haraway, Donna, 13, 133, 142, 146
Herberger, Sepp, 69–70
Hercher, Laura, 141
Herr, Hugh, 115, 120, 121
history and tradition of sport: disabled athletes, 125–26, 144; narratives of, 3–4, 15, 16–17, 99; sex and gender identity questions, 24, 144–45
Hobbes, Thomas, 12, 145

Hoff, Katie, 51, 57
Horween leather basketballs, 81, 82
hour record (bicycling), 20, 21, 86, 87–88,
  89–93, 175
Howman, David, 196–97
human body. *See* authenticity-body
  relationship; body narrative; mechanistic
  view of the body
human growth hormone (HGH), 3, 25, 154,
  201
Hungarian soccer team, 68–69, 70, 71

indirect testing for performance-enhancing
  substances. *See* athlete biological
  passports
injury repair, 104–5, 126, 211–12
instant replay technology, 5, 10, 212
International Association of Athletics
  Federations (IAAF): anti-doping efforts,
  26, 180–81; false start regulations, 213–14;
  Oscar Pistorius analysis, 21–22, 107–8,
  110–14; Oscar Pistorius appeal to CAS,
  112, 114–18, 119–120; Markus Rehm case,
  106; Caster Semenya case, 23–24, 135,
  136–140, 142, 143; sex and gender
  verification, 141–42, 146; track and field
  regulations, 132
International Olympic Committee (IOC):
  Lance Armstrong case, 156–57; Austrian
  team search, 189; ban on blood boosting,
  186, 191; Lausanne Declaration on Doping
  in Sport, 192–93; performance-enhancing
  substances investigations, 160; Oscar
  Pistorius case, 108; sex and gender
  verification, 24, 141, 146–49; testing of
  athletes, 23, 134; WADA creation, 191, 192
International Paralympic Committee (IPC),
  109–10
intersex athletes, 148, 149

Jackson, Mark, 82, 83, 84
Jackson, Stuart, 75, 81, 84, 85
Jaked, 59, 62, 63
Johnson, Jack, 134, 206
Johnson, Michael, 102, 126
Jones, Marion, 162, 168, 169, 171, 221
Jordan, Michael, 72–74, 206
Jordan-Young, Rebecca, 148, 149

Karkazis, Katrina, 148, 149
Kerr, Steve, 82, 84

Kimmage, Paul, 165
Kitajima, Kosuke, 58
Kram, Rodger, 115, 120, 121

Landeweerd, Laurens, 127
Landis, Floyd, 157, 172
Lane, Laraine, 137–38
Larfaoui, Mustapha, 61
Lausanne Declaration on Doping in Sport,
  192–93
Leveaux, Amaury, 63–64
Locatelli, Elio, 112, 116–17, 120
Lochte, Ryan, 51, 57
Longman, Jeré, 111, 151
Lugano Charter, 88, 89, 91
LZR swimsuits (Speedo), 19, 31, 48–53,
  56–59

machines. *See* mechanistic view of the body;
  motor-over-machine dialectic
MacRae Knitting Mills Ltd., 34–35
Magdalinski, Tara, 17
Major League Baseball (MLB), 25, 178,
  210–11. *See also* baseball
Malehopo, Molatelo, 137
Månson, Jan-Anders, 61, 62
Maponyane, Kakata, 137, 139
marathon race, 205
Marti, Jose ("Pepe"), 154, 244n20
Martin, Casey, 105
masculinity: in female athletes, 143, 150;
  sport cultural narrative of, 152
materials. *See specific types*
Mayer, Walter, 189
McGowan, Craig, 115, 121
McGraw, Muffet, 150
mechanical doping, as term, 95, 97–98
mechanistic view of the body: history
  of, 12, 145; prostheses questions, 126;
  in swimming, 51, 64; tension with
  human qualities, 13, 145–46, 221.
  See also cyborg athletes; motor-over-
  machine dialectic
Merckx, Eddy: hour record, 20, 21, 87–88,
  89–93; performance-enhancing
  substance use, 92; status as greatest
  cyclist, 20, 163
microdosing, 26, 166–67, 198, 199
Miller, Reggie, 84
Mizuno, 37, 38, 41, 58, 62
Mlangeni-Tsholetsane, Phiwe, 137

Mongel, Aurore, 63–64
Moore, Isabella ("Belle"), 33–34, *34*
Moser, Francesco, 20, 87
motor-over-machine dialectic: author's experience, 1–2; baseball bat materials, 219–220; dual effects of performance-enhancing substances, 209–10; multiple functions of, 4–5; need for symbiotic relationship, 14, 221–22. *See also* body narrative
Musciacchio, Giuseppe, 54

Naito, Haruo, 197
Nash, Steve, 84
National Basketball Association (NBA): Athletic Propulsion Labs shoe ban, 20, 74–77, 98; Nike shoe controversy, 72–74; synthetic vs. leather balls, 20, 80–86, 98
National Collegiate Athletic Association (NCAA), 217, 218, 253n29
net advantage concept, 22, 119–121
New England Patriots, 86
Nihon Suiei Renmei, 58
Nike: athlete sponsorship, 58, 173–75; prostheses technical assistance, 123, *124*; sports products, 11, 19, 59, 72–74, 81, 102
nylon materials, swimsuits of, 35

Obree, Graeme, 175–76
Oliveira, Alan, 121–24
Olympic Games: blood boosting concerns, 189, 191; fast suit use, 31, 37, 38, 41, 42–43, 46–48, 50; first women athletes, 33, 146; LZR Racer performance, 31, 52, 56–59; marathon race, 205; Jesse Owens participation, 20; Michael Phelps gold medals, 9; Oscar Pistorius bids, 21, 110, 118–19; Oscar Pistorius participation, 100–101, *101*, 118–19, 121, *126*; Russian doping, 160; Caster Semenya participation, 24, 131, 142, 151; sex and gender verification, 146–49; as showcase for technoscientific ingenuity, 9; swimsuit technology importance, 33–37, *34*, 65–66; testing of athletes, 134
omertà, in cycling, 157
Össur Flex-Foot Cheetah prostheses, 21–22, 108, 114, 117–18, 119–120, 124
Owens, Jesse, 16, 17, 20, 69, 234n1

paniagua, as term, 175, 246n59
Paralympics: 100 m race record, 108–9; Brian Frasure participation, 108, 122; Oscar Pistorius participation, 21, 106, 107–8, 110, 121–24
Parker, Candace, 150
Paterno, Joe, 174, 175
performance-enhancing substances: evolution of anti-doping movement, 2–3, 24–26, 221; list of banned substances, 182, 186, 192; mechanical doping vs., 95, 97–98; microdosing, 26, 166–67; prevalence of, 160–61, 179, 196–97, 209–11; technological doping vs., 19, 41, 52. *See also* athlete biological passports; *specific athletes*; *specific substances*
Phelps, Michael, 9, 31, 47–48, 51, *51*, 57–58
Piccione, Elisa Cusma, 143
Pistorius, Oscar: able-bodied events participation, 110, 111–12; CAS rulings, 22, 112, 114–18, 119–120, 124, 126; concerns about unfair advantage, 102–4, 106, 120–21; horse racing, 234n1; IAAF rulings, 21–22, 107–8, 110–14; Olympic bids, 21–22, 103–4, 118–19; Olympic Games race, 100–101, *101*, 121, *124*, 126; Paralympics participation, 106, 107–8, 110, 121–24; prostheses fitting, 108–10; questions of augmentation vs. assistance, 124–28, 144
Polara golf ball, 79–80
polyurethane swimsuits, 19, 50, 59, 60, 64
Portas, Cristiano, 53–54
positive drug tests: as aberrations, 210; burden of proof with, 181; false positives, 166, 168; reaction to, 171
privacy concerns, 139, 146, 170
prostheses: fitting of, 108, 123; Markus Rehm case, 106; sport authenticity questions, 106, 124–28, 212. *See also* Pistorius, Oscar
prostheses as unfair advantage: disabled vs. superabled arguments, 127–28; IAAF rulings, 21–22, 110–11, 112, 116; Michael Johnson concerns, 102; natural vs. unnatural arguments, 103; net advantage concept, 119–120; Markus Rehm case, 106; running tall concerns, 109, 110, 122–24

Puma, 11, 17, 69, 230n5
Pursley, Dennis, 60
Puskás, Ferenc, 68, 71

racial concerns: Cathy Freeman, 102; Jack
    Johnson, 134, 206
Radcliffe, Paula, 151
Rahn, Helmut, 69, 70
riblets, 40, 41, 227n20
Riccò, Riccardo, 195, 196
Rintala, Jan, 13, 145
Roche Pharmaceuticals, 195, 196
Rodriguez, Alex, 178, 211
Rogers, Neal, 172
rubber swimsuits, 36
Rubin, Stephen, 54
Rubio, Michael, 159
running, 16–17. See also track and field
running shoes, 19–20, 111
running tall, 109, 110, 122–24
Rushall, Brent, 42, 43
Russia, doping concerns, 160

Schubert, Mark, 57
Semenya, Caster, 136; early running career
    of, 132; inconclusive nature of reinstate-
    ment to competition, 142–44; Olympic
    Games participation, 131, 151; sex and
    gender verification questioning, 23–24,
    132, 135–140, 141, 148
sex and gender identity, 23
sex and gender verification: body-based vs.
    gender-based competition, 152–53;
    concerns with androgens as determina-
    tion, 141–42; embracing postgender
    competition, 144–45, 152–53; evolution of,
    23–24; historical testing, 146; IOC
    positions, 146–49; move to case-by-case
    assessment, 141, 146; overview, 131–35;
    physiological sex differences at elite
    levels, 133, 149–151, 152; purposes of,
    131–32, 151–52. See also Semenya, Caster
sex vs. gender, 133
Shimange, Oscar, 137
shoes: for basketball, 20, 72–78, 98; for
    running, 19–20, 111; for soccer, 68–72, 70
silk, swimsuits of, 34
soccer: ABPs implementation, 181; EPO use,
    178; goal line technology, 5, 213; Jabulani
    ball complaints, 79; shoes for, 68–72, 70;
    Women's World Cup, 151

Sottas, Pierre-Edouard, 26, 193–94,
    199
Spalding, 80, 81, 82, 84, 85
spandex swimsuits, 36–37
Sparks, Sam, 159, 160
Speedo: approval under revised swimsuit
    regulations, 62; competitor complaints
    against, 53–55; early company history,
    34–35, 226n3; Fastskin swimsuits, 41, 42,
    44–46, 47–48, 224n31; first fast suits, 37,
    38, 40–41; LZR Racer development, 19,
    48–53; marketing of fast suits, 42, 43,
    50–51, 51; spandex swimsuit develop-
    ment, 36, 37
Speirs, Annie, 33–34, 34
Spike Pad (Nike), 124
Spitz, Mark, 9
Stackhouse, Jerry, 84–85
starting blocks, electronic, 213–14
Steer, Irene, 33–34, 34
Stern, David, 72, 81, 83, 85
steroids. See anabolic agents
stride frequency, 120
stride length, 109, 121–24, 235n23
Struckhoff, Mary, 76
superabled athletes, 127
swimming: ABPs implementation, 181;
    recent technoscientific changes, 9;
    shattering of records, 32, 53, 58, 59–60,
    64, 65–66. See also Fédération Internatio-
    nale de Natation; swimsuit technology
swimsuit technology, 31–66; collaboration
    between manufacturers and technoscien-
    tific institutions, 48–49; concerns from
    Speedo competitors, 52, 53–55; early
    criticism of fast suits, 41–43; fast suit
    engineering, 37–41; FINA responses,
    55–56, 59–64; LZR Racer concerns, 31,
    48–53, 51, 56–59; in modern Olympics,
    33–37, 34; overview, 31–33; polyurethane
    suits, 19, 50, 59, 60, 64; reengineering of
    fast suits, 46–48; return to primacy of the
    body, 32, 64–66; scientific studies on fast
    suits, 43–46, 48
synthetic materials: basketballs, 20, 80–86,
    98; development of, 36; tracks, 17. See also
    specific materials

technological doping, as term, 19, 41,
    52
technological fraud, in cycling, 96

technoscience, as term, 2
technoscientific actors: competition among, 132, 208, 220; as constituency, 9–12; narratives of, 15, 67–68; sport authenticity questions, 215–16; symbiotic relationship with athletes, 11; uneasy equilibrium with other constituencies, 12. *See also specific organizations; specific sports*
testing: for fitness to compete, 134; as form of societal evaluation, 133–34; as truth making, 177, 180. *See also* direct testing for performance-enhancing substances; sex and gender verification
testosterone, 148, 154
textiles. *See specific types*
Thorpe, Ian, 39–40, 41
Timmerer, Julia, 96
Torres, Dara, 51
Tour de France: ABPs implementation, 195; Lance Armstrong wins, 163, *164*, 165; electric assist concerns, 95; Festina affair, 155, 187–88; French senate report, 188–89; as marker of excellence, 86; Eddy Merckx wins, 87
Tour of Flanders (Ronde van Vlaanderen), 93, 94
Toussaint, Huub, 44, 45
track and field: 100 m race records, 108–9; 400 m split times, 112; difficulties in historical comparisons of performance, 16–17; false starts, 213–15; gendered body tradition, 132, 133; performance-enhancing substances use in, 162, 180–81; sex and gender verification, 141–42. *See also* Bolt, Usain; Pistorius, Oscar; Semenya, Caster
training methods: altitude training, 190, 202; in swimming, 36; technoscientific approach to, 17, 184–85
Tucker, Ross, 122, 123, 124, 142
Tygart, Travis, 159, 168
TYR, 47, 62, 63–64

ulnar collateral ligament (UCL) reconstruction, 104
unfair advantage: assistive devices, 102, 103, 105; basketball shoe ban, 75; golf carts, 105; net advantage concept, 119–120; sex and gender verification, 137, 139, 141, 142. *See also* cheating; prostheses as unfair advantage

Union Cycliste Internationale (UCI): ABPs development, 178–79, 180, 181–82, 194–95; ABPs future plans, 201–4; ABPs implementation, 198–99, 201; Lance Armstrong investigation, 156–57, 162, 166, 174, 175; claims of clean sport, 210; hematocrit level testing, 186–87; history of, 178–79; hour record book revisions, 91, 92; response to electric assist concerns, 94, 95; revised regulations, 20–21, 88–89, 96, 98; sublimation of bicycles to bodies, 20–21; support for indirect testing methods, 26
United States Anti-Doping Agency (USADA): Lance Armstrong investigation, 25–26, 154–55, 156–160, 161, 166, 171–73, 175; Marion Jones investigation, 168; lax testing methods, 167–68; taxpayer support of, 159
United States Golf Association (USGA), 67, 80

Van Der Watt, Francois, 108
van Hilvoorde, Ivo, 127
vertical leap, 74, 75, 76
Vick, Michael, 174, 175
Vilain, Eric, 148
Vizard, Frank, 47
Voet, Willy, 155, 187

Wade, Dwyane, 82
Weiss, Pierre, 118
West German soccer team, 68–72
Weston, Mary Edith Louise, 146
Weyand, Peter, 115, 120, 121
whereabouts rule, 170, 183
White, Howard ("H"), 73
Whitehead, Mark, 191
Wielgus, Chuck, 64
Wilson-Roberts, Guy, 152
Winckler, Ciro, 123
women athletes: fast suit performance and body fat, 45; infrastructural development for, 150–51; Olympics participation, 33, 146; sex and gender verification, 24, 142, 146–49; small physiological differences with males at elite levels, 149–51. *See also specific athletes*
Women's National Basketball League (WNBA), 81

wood equipment, 8–9, 217
Woods, Tiger, 174, 175
Woodward, Ian, 15
wool materials, 8, 9, 32, 34, 35
World Anti-Doping Agency (WADA): ABPs development, 179, 180, 182, 193–95, 199–200, 201–4; Lance Armstrong investigation, 156–57; Austrian team search, 189; creation of, 191, 192; Haematological Working Group, 26, 193–94; IFPMA Joint Declaration, 196–97; move away from direct testing, 199–201; removal of caffeine from banned substances list, 203, 211; report on Russian doping, 160; response to Daly report, 167; rise of, 191–95; testing of athletes, 23, 134; whereabouts rule, 170, 183
World Conference on Doping in Sport (1999), 192–93
World Cup: goal line technology, 213; Jabulani soccer ball complaints, 79; shoes for, 68–72; Women's World Cup, 151

Y chromosome, as indicator of sex, 141

Zettler, Patricia, 125
Ziegler, Kate, 51, 57